CASE STUDIES IN CONSTRUCTION PROJECT MANAGEMENT

CASE STUDIES IN CONSTRUCTION PROJECT MANAGEMENT

KB RAJORIA & HK SRIVASTAVA

KHANNA BOOK PUBLISHING CO. (P) LTD.

Publisher of Engineering and Computer Books

4C/4344, Ansari Road, Darya Ganj, New Delhi-110002
Phone : 011-23244447-48 **Mobile:** +91-9910909320
E-mail : contact@khannabooks.com
Website : www.khannabooks.com

Case Studies in Construction Project Management
by **KB Rajoria & HK Srivastava**

ISBN: 978-93-55389-36-7

First Edition: 2025

To view complete list of
books, please scan the
QR Code:

Published by:
Khanna Book Publishing Co. (P) Ltd.
CIN: U22110DL1998PTC095547

Visit us at: www.khannabooks.com
Write us at: contact@khannabooks.com

भारत सरकार
Government of India

CPWD

ailendra Sharma
Director General

निर्माण भवन, नई दिल्ली-110011
Central Public Works Department
Nirman Bhawan, New Delhi-110011
Tel : 23062556/1317, Fax . 23061884
E-mail : cpwd dgw@nic.in

Foreword

While construction management of projects focuses on their engineering aspects, contract management is inclined more towards their legal aspects. Project Management is a science, which takes into account both of these aspects, and ensures that the projects are planned, executed and closed within the scheduled cost, quality and time. Project management ensures that all stakeholders, including the contractor, are engaged to get timely action on engineering or contractual issues. This documentation of case studies on construction projects will be an invaluable resource to all engineers, managers and contractors.

Majority of the case studies included in this Compilation reflect personal experiences of the authors. A few case studies of projects executed by other colleagues of CPWD and PWD Delhi are also included to cover the desired variety of projects. Case studies of other organizations included in this compilation are also experiences of CPWD cadre Engineers, posted to these organizations on deputation. These organizations include Water and Power Development Consultancy Services (WAPCOS), National Building Construction Corporation (NBCC), Delhi Development Authority (DDA), Delhi Tourism and Transport Development Corporation (DTTDC) and Andaman PWD (APWD). Engineers and Administrations of these organizations also deserve appreciation for their contribution to the projects.

The scope, environment, location, timelines, administrative set up for each and every case study is different and brings in different perspectives. The successful execution of the projects through good project management speaks high of the engineers who worked hard and with sincerity on these projects. Out of these cases, Padma Shri has been awarded to respective Chief Engineers for honorable completion of two projects. Shri Balbir Singh Saigal was awarded the Padma Shri for Siddharth Raj Marg in Nepal and Shri O P Mittal was awarded the Padma Shri for Parliament Secretariat Building at New Delhi.

Another shining example was when the works executed by CPWD were highly appreciated by the Government of Arunachal Pradesh and awarded gold medals, silver medals and Commendation Certificates. My compliments to the project team for their sincere hard work, service to the organization and ultimately, service to the society.

I sincerely hope that this repository of experiences will be found interesting and useful by engineering professionals and project management personnel. I wish that engineering colleges, IITs and other technical institutions include such course material in their syllabus for graduate and post graduate studies so that the students may get the flavour of field work besides their other academic pursuits

My compliments to authors and other professionals who worked towards the documentation and publishing of these Case Studies.

(Shailendra Sharma)

PREFACE

Construction is one of the most important activities for the growth and development of any country. Once a Construction project is conceived and identified, it is an earnest desire of all concerned that all activities to implement the project work are performed expeditiously and effectively. Important activities include identifying land, conceiving the project, preliminary planning and approval, financial approvals, availability of services and approval from statutory authorities. The implementation process includes many activities such as detailed planning, drawings, estimation, approvals, etc. The next stage includes deciding on contracting agencies, supervision agencies, testing, checking, quality assurance and acceptance. Finally, the project is cleared for intended use by the user for its design life, methodology, and maintenance norms.

Case studies of various projects in this book give comprehensive details about the contribution of engineers, planners, architects and others, from the conceiving to the completion stage and beyond. These studies also describe the role played by top-level engineers to ensure the successful completion of projects. These studies illustrate how engineering professionals took political and administrative support and guidance, to remove hurdles and obstacles and ensure that the project was implemented in a hassle-free manner within the stipulated time. To implement the project, the concerned Secretary, Chief Secretary, Minister in charge, Chief Minister & even the Lieutenant Governor play an important role.

The technology for construction has been changing rapidly, and the adoption of new technology is necessary. A change is considered desirable according to the requirements of a project to ensure the use of more efficient technologies. After deciding to use any new technology, extra care during implementation is necessary. Project implementation becomes a challenging task when unrealistic time schedules are imposed. Engineers from the top-level to the junior level are required to work efficiently to ensure the timely implementation of the project. Working on projects in remote & rugged areas is another challenge. Even for these projects, timely financial approvals, innovation and hard work by engineers are necessary to ensure completion in minimum time. It is human courage that is supreme.

The specialty of civil construction is that each & every project is different. There is no repetition. Therefore, engineers always face challenges in accomplishing what they set out to do. Unique solutions have to be found for each & every situation in different projects. This challenge makes professionals creative, and they have to be optimistic in their approach. The engineering profession is, therefore, difficult and exciting. For effective & efficient project management, solutions are to be found at each & every stage of the project. Case studies in this book show that projects were completed because of proactive initiatives and out-of-the-box thinking by all concerned from top-level to junior level.

This book covers case studies from 1963 to 2020, almost sixty years. With time, technology, methodology and approach continue to change. Nevertheless, human behavior, interactive capabilities and desire to perform are consistent and will remain the same. Therefore, these case studies are relevant for today & tomorrow. Professionals can learn from these case studies. Students can learn lessons about the practice of construction management. It is the success that counts; how it can be achieved is the ingenuity of the team. The credit for the project's success goes to those who worked on these projects.

These case studies are documented by Central Public Works Department (CPWD) cadre engineers working in Central PWD and other organizations. It is well known that Central PWD is a premier Central Government agency in-charge of public sector works. Under the Ministry of Urban Development, Govt. of India, the CPWD deals with buildings, roads, bridges, flyovers, and other complicated structures, including stadiums, auditoriums, laboratories, bunkers, border fencing, and hill roads. It provides services from project conception to completion, and maintenance management. It is headed by the Director General (DG), the Principal Technical Advisor to the Government of India. Different public works departments and the Engineering Wing of several departments follow CPWD Specifications and Manuals.

Other organizations include, Delhi PWD, Delhi Development Authority (DDA), National Building Construction Corporation (NBCC), Delhi Tourism and Transportation Development Corporation (DTTDC), Andaman PWD, and Water and Power Development Consultancy Services (WAPCOS). It is considered desirable to introduce these organizations briefly.

(i) **Delhi PWD** - The Delhi PWD is headed by Engineer in Chief. The Delhi PWD works independently under the administrative and financial control of the Delhi Government except for the control of the cadre. All recruitment, promotions, transfers to and from the PWD to CPWD, quality assurance functions, and vigilance functions are carried out by CPWD.

(ii) **Delhi Development Authority (DDA)** – The Delhi Development Authority (DDA) is a planning authority created in 1957 under the provisions of the Delhi Development Act. It is responsible for planning, developing and constructing housing projects, commercial lands, land management and providing public facilities like roads, bridges, drains, sports centers etc., within the area of the National Capital Territory of Delhi. It is under the administrative control of the Central Government through the Lt. Governor of Delhi. For engineering works of DDA, there is a separate engineering cadre. From time to time, high-level posts of DDA like that of Engineer Member and Chief Engineer are handled by Central PWD Engineer.

(iii) **National Building Construction Corporation (NBCC (India) Ltd)** – It was founded in 1960 as a Govt. of India Civil Engineering Enterprise under the Ministry of Urban Development. NBCC has been involved in the execution of diversified projects in sectors such as Institutional, Housing, Industrial & Environmental, Transportation, Power, etc. NBCC ventured into Overseas Operations in the year 1977, executing projects of diverse nature in various countries. It has its Engineering cadre. This Ministry appoints the Chairman and Managing Director (CMD) of NBCC. Engineers of the Central PWD cadre have worked in NBCC, as CMD and on several higher posts like that of Chief Engineer and Superintending Engineer.

(iv) **Delhi Tourism and Transportation Development Corporation (DTTDC)** – This Corporation works under the administrative control of Govt. of Delhi. It was incorporated in 1975 to promote tourism in Delhi. When the mandate for the construction of flyovers was entrusted to the Corporation in 1989, its name was changed to Delhi Tourism and Transportation Development Corporation. Funds generated by this corporation are utilized to construct various Transportation Infrastructure Projects in Delhi. All the Engineers are taken on deputation basis from CPWD. The CPWD Manual is followed for the execution of works.

(v) **Public Work Department for Andaman and Nicobar Islands (APWD)**, was set up in 1957 and is under the administrative control of the Lt. Governor. It serves the people of these islands by creating infrastructure and providing comprehensive services including planning, designing, constructing and maintaining office and residential buildings of various departments. At present, APWD has its engineering cadre. Top posts at times are held by Engineers of Central PWD cadre.

(vi) **Water and Power Development Consultancy Services (WAPCOS Limited)** is a public sector undertaking of Govt. of India under the Ministry of Jal Shakti. The technical workforce is taken, as per requirement, on deputation. Required engineers are taken from Central Water and Power Research Station and Central PWD.

AUTHORS

K.B. RAJORIA

K B Rajoria did BE (Hons) in Civil Engineering from MBM Engineering College, Jodhpur in 1961 and ME (Gold Medallist), in Soil Mechanics and Foundation Engineering from Indian Institute of Technology Roorkee in 1963. He was also a faculty member in this Institute. He joined Central PWD in 1963, and worked in different capacities in the Government including hill road project in Nepal, prestigious Parliament Secretariat Building in New Delhi, several building and hill road projects in Arunachal Pradesh, Water Research Center Project at Baghdad (Iraq), projects for Delhi Development Authority etc. He worked at the Government of Delhi where projects to his credit include Delhi College of Engineering, Hospital Projects, Delhi Secretariat Building and a number of flyover projects. He superannuated as Engineer-in Chief Government of Delhi.

He was bestowed with important honorary positions such as President, Indian Buildings Congress, President, and Indian Roads Congress as also Chairman, Quality and Safety Forum, the Institution of Engineers. He was awarded Gold Medal for outstanding services in Arunachal Pradesh, Certificate of Outstanding Achievements by Govt. of Delhi and Award for 'Outstanding Engineer' by the Institution of Engineers.

He has published several technical papers in professional journals and authored books on various aspects of construction. A book titled; "ISO 9000 Practices in Construction" authored by him has been published recently.

H K SRIVASTAVA

H K Srivastava, did B Tech (Hons) in Civil Engineering in 1969, and Masters in Structures in 1971, both from IIT Kharagpur. He joined faculty at IIT, BHU for one year prior to joining CPWD in 1972. He worked in different capacities for more than 38 years from Assistant Executive Engineer to Additional DG, CPWD.

H K Srivastava has worked in different geographical areas as well as across different organizational cultures within the country and abroad. Worked on road projects in Arunachal Pradesh, East-West Highway Project Nepal, Delhi PWD, Indo Bangla Border Road Project and Rural Roads for PMGSY. He has carried out design of multi storied buildings at Delhi, as also bridges in Nepal and West Bengal. He worked as Chief Engineer Delhi Metro Rail Corporation in its initial phase. He has experience of working for Ministry of Finance in finding real estate investments for Income tax purposes, and as also with NBCC on projects in Iraq at Kirkuk, Mosul and Baghdad during mid-eighties.

He is a member of various professional bodies. He has a number of technical papers to his credit. As an active member to Human Resource Development Committee of IRC, he has been closely associated with the development of 'Guidelines on Skill development of Workmen in Road Sector (IRC 127-2018) and 'Guidelines on Training of Highway Professionals' (IRC: 128-2019).

INTRODUCTION

This documentation aims to share firsthand experience of the Engineers as to how different projects under different circumstances and environments were executed successfully. Chapters in this book have been arranged under two subheads - 'Built Environment Sector' and 'Roads and Runway Sector'. Authors considered it desirable to include a few contemporary projects also to highlight the uniqueness and complexities of these projects. The role played by individuals and teams to sort out difficult situations has been explained in each case. The management principles applied have been highlighted in the analysis part of each Chapter. It is expected to provide young architects and engineers insight into evolving innovative solutions in dealing with complex situations that could arise during construction projects planning and implementation. These projects were executed at different geographical locations in India such as Delhi, Faridabad, Lucknow, Arunachal Pradesh and Andamans. Projects of Nepal and Iraq have also been included in the book. Authors played important roles in many projects described in this book.

1. **Chapter 1** is about a Housing project of CPWD at Faridabad, Haryana. This Chapter describes the leadership role in the project completion stage, which includes resolving outstanding issues, reviewing incomplete work's status, identifying the means and ways to complete such works, preparations and payment of final bills, arbitration cases, and audit observations. Besides, after the completion of projects, issues regarding maintenances were resolved to the resident's satisfaction. This Chapter gives valuable insights into how a seasoned officer gave clear direction to sort different issues and personal guidance to his immediate junior officer to face challenging tasks ensuring the completion of projects.

2. **Chapter 2** is about the Capital of Arunachal Pradesh. It describes the success story of the construction of temporary capital for the newly formed Union Territory of Arunachal Pradesh. This Chapter describes the challenges undertaken for developing a new township and elaborates on various tasks for the completion of projects. There is no other example of such fast township development in our country. Gold Medals, Silver Medals and Commendation Certificates were awarded to Engineers for their services.

3. **Chapter 3** deals with development works at Pasighat, Arunachal Pradesh. It describes crucial projects executed during the early seventies at this remote and distant place of Arunachal Pradesh. Contractors were unavailable in the area during that period and the normal tendering procedure was ineffective. After

taking sanctions from the Government the projects related to college building & micro hydel projects were completed timely. One 'Inglish Bridge' bridge could be erected with the ingenuity of a young officer and with the help of a tribal contractor. Arunachal Pradesh Administration suitably awarded medals and commendation certificates to Engineers.

4. **Chapter 4** is about Parliament Secretariat Extension Building in Delhi. It describes the difficulties faced in the completion of the finishing and furnishing work of this building. The finishing and furnishing work of the project was to be completed in six months for an international event for which the date was already decided. There were interventions by high-level officers of the Parliament Secretariat, which were mutually sorted out. The timely completion was highly appreciated and Chief Engineer was awarded Padma Shri for his contribution.

5. **Chapter 5** deals with resettlement colonies in Delhi. It details the development of the biggest resettlement project ever executed in India. It was implemented by Delhi Development Authority (DDA). The total area under the scheme was about 100 hectares. During the implementation of the project, it came to notice that the Resettlement Colonies of the Trans-Jamuna area were likely to be subjected to flooding during rains. To sort out this difficulty, a Lake in an adjoining area was developed by excavation, and level of Resettlement Colonies raised by this earth. Later, norms and frequencies for maintenance activities for Resettlement Colonies were also developed for public convenience.

6. **Chapter 6** is about the development of the Water Research Centre Project in Bagdad Iraq. It describes the role played by Indian Government agencies for this project. The consultancy services were given to the Govt. of Iraq by Water and Power Consultancy Services with technical backup from Central Water and Power Research Centre (Pune) and Project Engineers from CPWD. It is a classic example of different entities working in a coordinated manner. A vital role was played by Engineering Projects India Ltd (EPIL) in the construction work during the hostilities with the neighboring country Iran. The project was completed after the Iraq- Iran war was over. For the completion of the project, funds were arranged from India by M/S EPIL. These funds were repaid later by the Govt. of Iraq to Govt. of India in commodities like oil. It has no parallel example.

7. **Chapter 7** is about construction of a twelve storied Hotel at Mosul, a very important city of Iraq during early to mid eighties. It was implemented by National Building Construction Corporation Ltd (NBCC), a Govt. of India undertaking. Covered area was 20,000 sqm. For this project Ziggurat type architecture was a main feature which was in tune with ancient Iraqi architecture of 2100 BC. It was a turnkey project. All the Civil works were completed

departmentally with workmen from India. All equipment and materials were imported. Special type of fittings and fixtures were used for this super deluxe hotel project. Providing finishing to the requirement in coordination with various disciplines was really challenging.

8. **Chapter 8** is about construction of Water Treatment Plant at Dibis (Iraq), by National Building Construction Corporation Ltd. Dibis was situated in North East Iraq. The Water treatment plant was developed for treating water for pumping to oil wells, so that oil moves up and could easily be pumped out. During that period, no construction material, equipment, spare parts were available in Iraq and had to be imported. Works for this Project were executed using departmental labour. Works of laying 2 m thick raft foundation for the water intake structure under dry conditions and providing staging for double height pump room were really challenging. Difficulties encountered and solutions found have been described in the write-up.

9. **Chapter 9** describes the development of services in regularized unauthorized colonies in Delhi. It was during 1984-85 that Delhi Development Authority started development activities for Regularized Unauthorized Colonies. It was a new, innovative and bold initiative to suit the reality of life. A number of these colonies were in East Delhi. The residents wanted a paved road in the first instant. However, on engineering consideration the drainage system of Trans Jamuna areas was improved in the first instant. After that, roads and drains were provided. By providing functional services in a designed manner, resident's quality of life in these colonies improved and provided long-term benefits.

10. **Chapter 10** gives details of the quality system developed for the Vasant Kunj Housing Project of DDA. It was a great responsibility to provide safe housing with the required standards. First, the specifications were decided and systems were developed to ensure stage checking of every building portion. Proper records were also kept. The next stage of construction was taken up only if a satisfactory report for the previous stage was available. Besides, processes were developed to ensure the checking of services for leakages before covering. The development and adoption of such a concept were path-breaking.

11. **Chapter 11** deals with strengthening of foundations for an under-construction Housing Project in Delhi. During early eighties, it was noted that in Kishangarh Housing Scheme of DDA, the size of foundations was smaller than required as per design. A workable solution to strengthen the foundation was framed in consultation with experts. The allottees, who were greatly concerned and agitated, were informed about the proposed engineering solution. The strengthening of the foundation was accordingly carried out. After that, allottees occupied these houses without reservations. Even after more than forty years, there is no complaint.

12. **Chapter 12** describes the construction of the OPD Block and Ward Block of LNJP Hospital Complex New Delhi. Construction of a new multistoried complex was getting deferred for years as existing buildings, though crumbling and unsafe, were entirely in use and could not be demolished. A unique solution was worked out, which was acceptable to all concerned. Brain storming sessions and detailed homework led to the construction of some essential services buildings in areas where no new buildings were to be constructed. Thus, dividing the problem into parts helped in finding workable solutions. A quick decision-making system was established by frequent visits of the Chief Engineer and his availability for problem-solving just a call away. Because of bold decisions taken by the Health Minister and Finance Minister of Govt. of Delhi the work could be implemented in record time.

13. **Chapter 13** describes comprehensive details of the construction of the Delhi Sachivalya Building. After Asiad 82, an RCC Framed structure was left incomplete in the compound of the Indoor Stadium Complex. Originally it was planned for the player's accommodation during Asiad 82 Games. Later this incomplete building was proposed to be used for a private hospital. They undertook some additional and alternative work. It did not proceed further due to various problems. Later on, it was given to Delhi Government for its Secretariat. After extensive study, PWD Engineers decided to strengthen RCC members for the structural safety of this building. It involved meticulous analysis and planning. After that some additions and alterations were undertaken to use it as an office. The finishing works were carried out to the appropriate specifications. It houses the Secretariat of Govt. of Delhi.

14. **Chapter 14** describes the Parliament Library Building constructed in the Parliament House Complex. It is Delhi's most important building project after the Central Secretariat Complex executed during the early twentieth century. The Architectural form of the building was conceptualized to achieve a low-key architectural expression signifying sagacity and spiritual elegance rather than to compete with the Parliament House. The height of the building was restricted to the podium level of Parliament House, except that glazed crystalline forms of domes protrude above the podium level. It was constructed following ISO 9000. State-of-the-art techniques and technologies were used for electrical and air conditioning work. There is no building of equitable standard in Delhi

15. **Chapter 15** is about the Development of the Delhi Technological University Complex Building. It describes how the Delhi College of Engineering was started at Kashmere Gate Delhi in 1941. The activities continued expanding, and the need for more space was projected. In 1980, the land was identified in North West Delhi near Bawana Road. This Chapter describes the construction of a new complex. It was low-lying land to suit the building complex. A private Architectural firm did the planning work. The whole complex was

constructed with a horizontal spread for different functional buildings. There were many difficulties in building construction due to poor soil conditions. Innovative solutions were found to overcome difficulties. The whole complex was completed by 2000 and the Delhi Collage of Engineering was converted into Delhi Technological University. It was one of the most challenging building project complexes taken up by PWD Delhi.

16. **Chapter 16** describes a building project taken up for Accountant General's office in Lucknow. Unique solution sorted out difficulties concerning foundation design. The quality standards of high order were achieved. Contractual problems were sorted out with a proper assessment of the situation and work completed well within sanctioned cost. The client highly appreciated the work.

17. **Chapter 17** describes the Maintenance of CPWD Buildings in Delhi. It deals with issues connected with the maintenance of government buildings. After completion of construction, buildings are required to be kept in proper condition when in use. Several innovative initiatives were taken for improvements in maintenance which have been described. It was noted that specifications must be improved during the maintenance period, as per specific needs. With proper management and empowering of field staff, results were spectacular. Efforts were made to recondition and reuse rather than replace them with new substitutes. This kept the workers engaged in different activities, resulting in greater job satisfaction and economy. The end users (occupants) were treated as customers and their satisfaction was considered of prime importance.

18. **Chapter 18** describes the Construction of the 180 km long Siddarth Raj Marg in Nepal during the mid-sixties. It was the first project executed by Central PWD in a foreign country. The project was completed according to the schedule. The strategy for the execution of the project was stationing Executive Engineers along the road alignment. The Chief Engineer played the most important role by visiting the road alignment on foot. Difficulties were recognized, ascertained, appreciated, and sorted out then and there. Another significant contribution was by Superintending Engineer, who decided on geometric standards and specifications. Executive Engineers with Superintending Engineers, implemented the project by finding innovative solutions to problems encountered. The Chief Engineer was awarded Padma Shri for his meritorious services.

19. **Chapter 19** gives details of innovative initiatives taken to improve Pasighat-Pangin Road in Arunachal Pradesh it to make it an all-weather road. This road was the only link for connecting interiors with Pasighat. The administration was keen to connect with the local population living in the Interiors. It became possible with all-weather connectivity. Initiative taken included cutting along small river banks to make stable log bridges, which could stand during rainy season and making a timber bridge of 50 feet span by transporting tree trunks

with the help of elephants. Besides, road work was started with proper geometry, hill drains, culverts etc., both from Pasighat and Pangin sides. Practical instructions were given to the field engineers for achieving geometric standards. The Local population was happy with the improvements on this road.

20. **Chapter 20** is Yingkiyong Damroh Road Arunachal Pradesh. It is about the construction of Yingkiyong Damroh road in the deep interiors of Pasighat Sub-division of Arunachal Pradesh Resources was mobilized with the help of local tribal leaders, village heads, educated youths and local contractors. Arrangements were made to provide living facilities near the work site. The Administration did an arrangement of food items for the workers. A favorable environment was created for work to proceed unhindered. Delegation of authority to the field engineers helped in the speedy execution of work. The manner of implementation was worked out according to the local environment. The project was completed in a record time of one and a half years and was highly appreciated by Arunachal Administration. Engineers and Local Leaders were suitably awarded.

21. **Chapter 21** is about ISBT Flyover and Yudhishthir Setu Across River Yamuna, Delhi. It was one of the most important projects undertaken at that time. This was the most congested location in Delhi during that period and even crossing the river Yamuna was difficult preposition. The curved flyover near ISBT Delhi, provided a unique signal-free solution in two levels to solve traffic problems of that location and the area around the bus terminus. Completion involved solving contractual problems with out-of-the-box solutions.

22. **Chapter 22** deals with Loni Road Flyover in East Delhi. It describes the project undertaken at this location with heavy traffic congestion. A Casting Yard was developed for pre-cast girders near Bhatti Mines of Delhi. These girders were transported to the project site. Precast girders were provided on approaches to main flyover except for central cast in situ span. This technique helped in expeditious of construction and also better control of quality. After that, several two-level flyovers were constructed with this technology on Outer Ring Road.

23. **Chapter 23** describes the planning and construction of the Punjabi Bagh Flyover- at crossing of Ring Road and Rohtak Road. It is the only four-level fly-over in the country. Such a simple but bold concept, especially at a busy intersection in an urban area, required special techniques and measures. Using a diaphragm wall as a substitute for a conventional retaining wall avoided problems associated with deep excavation at a busy intersection. In order to balance the upward pressure of ground water, iron ore filling was done for the underpass road. To ensure traffic flow along Ring Road, a temporary steal bridge was constructed at the location of the Flyover. It was also used as a support structure for constructing the Main Flyover.

24. **Chapter 24** deals with the planning and construction of Flyover projects at Dhaula Kuan and AIIMS, Delhi. Delhi Urban Arts Commission did not approve the proposals prepared by PWD. With intervention by Chief Minister, an Empowered Committee was constituted and acceptable solutions were found. An open press notice was given for the concept designs of these Flyovers. Designs given by different persons were evaluated and the Empowered Committee selected the best proposals given. At Dhaula Kuan, due to the difficulty of land from the Ministry of Defense, designs had to be changed to complete the project. Both these flyovers are unique examples of resolving of administrative hindrances as and when they arose and evolving new design concepts.

25. **Chapter 25** gives details of Flyovers with Precast Segmental Technology on Ring Road and Outer Ring Road in Delhi. It was a courageous initiative that involved risk, as technology was earlier not tried in the country. It evolved through competitive design concepts invited by structural consultants. The mandate was to select a design that involved minimum construction activities at the site, as no open space near the busy intersections was available for traffic diversion during construction and diverting affected utilities. Four Flyovers constructed with this technology are at Ring Road crossings near Moti Bagh and Safdarjung Enclave and Outer Ring Road crossings near Savitri Cinema and Nehru Place. A Casting Yard for the segments was set up near Nizamuddin Bridge. The quality standards were based on ISO: 9000. It was the confidence of engineers and consultants that the projects were completed without any difficulty.

26. **Chapter 26** deals with Flyovers using Steel Fabrication Technology on the Ring Road in Delhi. The unique solution adopted to save time in construction has been described. These flyovers were constructed on Ring Road intersections at Khel Gaon Marg and Kirti Nagar. The bridge components must be fabricated in manageable sizes for ease of transportation. Thus, fabrication skills and quality assurance were of utmost importance. Components for these flyovers were fabricated at L&T Workshop, Chennai, under controlled conditions. After transportation, steel members were joined by bolting. Special long-lasting paint was used to reduce maintenance costs and increase durability.

27. **Chapter 27** describes the steps adopted in using fly ash for embankment construction in the approach road to the Nizamuddin Bridge constructed with Japanese aid. For the first time, a considerable quantity of fly ash was used to embank a river bridge. The design was based on specifications framed with the technical studies conducted by the Central Road Research Institute (CRRI) Delhi. As the fly ash was being used for the first time for a river embankment, DG Roads did not approve the proposal. It became an issue of difference in opinion between State Government and Central Government and a meeting was held in the office of the Cabinet Secretary. After considering the basis for selection and likely impact. it was decided to use fly ash for this project. Later,

IRC also accepted the use of fly ash in river embankments.

28. **Chapter 28** deals with the Development of Port Blair Airport, Andaman and Nicobar Island. First, the length of the existing runway was to be increased. At the project's implementation stage, the requirement was upgraded so that the airport could become suitable for taking the load of heavier aircraft. During the execution stage, there were difficulties, primarily due to inadequate planning by consultants in project preparation. The project's status was reviewed and modified by Engineers with the approval of the Andaman Administration and a revised estimate was submitted to the Ministry of Civil Aviation. As the Airport was primarily used for defense purposes, the working hours were limited between 5 pm to 5 am only. Innovative solutions were found, and the airport project was completed under challenging conditions. Decision-making for the smooth project implementation was done by a high-power group headed by Lt. Governor. Under the leadership of the Chief Engineer, the project was completed smoothly.

29. **Chapter 29** describes how underpass at Madhuban Chowk was constructed with prestressed soil anchors to act as a stabilizer against uplift pressure on the foundation for the underpass. This technology was used for the first time in the country. It was impossible to construct a flyover at this place, because of the overhead Metro line. This project was challenging and considerable planning was done to implement the new technology.

30. **Chapter 30** describes the construction of the landmark Signature Bridge on River Yamuna. A decision was taken at the Delhi Government level to construct a cable-stayed bridge with Central Pylon having about 150 m height (double the height of Qutub Minar). For the planning, design, and construction of this bridge, experts from several countries participated. Besides, for fabrication works, a workshop in China was selected. The transportation of fabricated parts was also a complex and unique proposition. Indian Engineers and Contractors constructed the bridge and it approaches. Signature Bridge at Delhi stands as an unique example of excellence in engineering.

GRATITUDE

1. Authors are honored to be a part of amazing CPWD Organization providing multiple opportunities in challenging projects and be associated with people who have passed on many useful lessons for success in professional as well as personal life. Each and every situation - good and not so good, project, persons gave opportunity to learn and transfer some lessons. Our thanks to them for being the inspiration and foundation for the leadership manifesto. Without the guidance and support of peers, this book would not have come to the present shape. Shri O P Goel, former Director General CPWD, provided encouragement and guidance for a number of chapters of this compilation. Our regards and gratitude to him. Our gratitude is to Shri M K Agarwal, former Engineer-in-Chief PWD Haryana, an eminent Civil Engineer, for providing detailed template for the narrative. He also encouraged authors for preparation of such a document. Shri Deepak Gupta, Consultant Management System Group, reviewed all the case studies and contributed by making suitable suggestions to make the compilation very purposeful for the readers. Editorial help provided by him is also gratefully acknowledged. Shri Murali Jagannathan, from National Institute of Construction Management and Research(NICMAR), provided crisp comments and suggestions on each of the chapters which have added value to the content of chapters. Shri Vijay Verma former Engineer-in-Chief PWD Madhya Pradesh guided us in adding important aspects in the writeup relating to the theme of the book. Our gratitude to him. Help and support from peers has been very useful.

2. As the work progressed, it was considered necessary to add varied examples. This prompted the authors to invite a few other authors to share their experiences as also by adding contemporary projects for which papers were published by different authors.We are grateful to all the contributors. For sake of avoiding multiple repetitions, senior most position held by the contributors have been indicated, without prefixing 'retired'.

3. For Chapter 2 Capital of Arunachal Pradesh, Shri H K L Mehta, Additional DG, added very useful information. For Chapter 4 Parliament Secretariat Extension Building, Shri Ashok Agarwal, Chief Engineer, assisted in compiling information. For Chapter 5 Resettlement Colonies, guidance was given by Shri R G Gupta, former Commissioner Planning DDA, to make the document purposeful. Chapter 7 is for Water Research Centre. Shri R K Gupta of EPI and Shri S K Singhal, Chief Engineer, helped in drafting of the paper. For Chapter 2 Capital of Arunachal Pradesh as also Chapters 3, 19 and 20 relating to works of Arunachal Pradesh, Shri A K Sarin, Additional DG, not only provided useful information but also assisted in adding first hand description of important activities.

4. Basic data, encouragement and guidance for Chapter 5 'Resettlement colonies of Delhi' was made available by Shri R A Khemani, Chief Engineer DDA. Chapter 15 Delhi Technological University was prepared by Shri P S Chaddha, Additional DG, with inputs from Shri B K Chugh DG. Both were Projects managers during implementation. Chapter 16 Central Govt. Office Building at Lucknow was contributed by Shri Ramesh Chandra, Chief Engineer. Chapter 28, Extensions & Strengthening of Runway at Port Blair Airport, Andamans and Nicobar Islands was authored by Shri H S Doga, DG.

5. Technical publications have been the basis for following chapters. Authors express their gratitude to authors of these technical papers. Chapter 13, Secretariat Building for Govt. of Delhi the publication of Shri DS Sachdev, DG; Chapter 14 Parliament Library Building two compilations of activities covering civil works including service and Electrical, Electronic and Mechanical services have been referred. These compilations were published under the leadership of S/S Jag Mohan Lal, Additional DG, JN Bhavani Prasad, DG and K Srinivasan, DG; Chapter 21, ISBT Flyover and Yudhister Setu across river Yamuna, Delhi, paper of Indian Roads Congress authored by S/Shri LR Gupta DG, A Chakrabarty DG and SS Mondal DG; Chapter 22, Loni Road Flyover with Precast Girders, help was taken from Indian Roads Congress Paper no. 416 An Innovative Approach for Urban Flyover Construction, by S/Shri PB Vijay DG, A Chakrabarty DG and S S Mondal DG. Further inputs were provided by Shri A K Bajaj Additional DG;Chapter 23, Punjabi Bagh Flyover Paper "Multilevel Grade Separator at Ring Road – Rohtak Road Intersection at Punjabi Bagh, Delhi by Deepak Narayan Additional DG, KN Agrawal DG, B K Chugh DG and S K Rustagi Special DG; Construction of Grade Separators at the intersection of Ring Road and Aurobindo Marg New Delhi – A Success Story, by S P Banwait Additional DG, S S Mondal DG and Rajeev Singhala Chief Engineer in IRC Journal for Chapter 24; Chapter 29, Underpass at Madhuban Chowk, information has been taken from article "Use of Prestressed Soil Anchors in Construction of an Underpass in High Water Table Zone" prepared by Shri Shishir Bansal (presently working as) Chief Engineer, V Gupta Additional DG and Jose Kurien Additional DG published by Federation Internationale du Beton (FIB) in proceeding of 2nd International Congress June 5-6, 2006, Italy; Chapter 30, Signature Bridge Delhi, has been abstracted from the paper Bridge on River Yamuna and Iconic Structure published by Shri Shishir Bansal in the Institution of Engineers in their Civil Engineering Journal 2021. Further inputs for both of these Chapters from Shri Shishir Bansal along with photographs is gratefully acknowledged. The authors also acknowledge documentation carried out by Shri S P Banwait Additional DG in respect of 'Flyovers with Steel Fabrication Technology on Ring Road', Chapter 26.

6. Photographs have been taken from different sources, besides photographs available with authors in their own record. Satellite pictures have been taken from Google/

Google Earth. Roof top view of Parliament Library is taken from M/S Raj Rewal Photo. All the sources are gratefully acknowledged. Besides for a number of flyover projects and Delhi Secretariat Building photographs have been taken from PWD Delhi for which we are grateful to Shri Anant Kumar, Engineer-in-Chief, Shri P K Parmar Chief Engineer and Shri A K Gupta Executive Engineer. For CAG Building at Lucknow, Parliament Library Building and a few of the flyover projects at Delhi, photographs have been provided by Shri Ramesh Chandra, Chief Engineer. Shri K N Agarwal, Director General, Shri S K Rustagi, Special Director General, provided photographs for Punjabi Bagh Flyover, AIIMS Flyover and Dhaula Kuan Flyover. Besides photographs for Delhi Technological University was provided by Shri PS Chadha, Additional DG and Vidushi Rajoria. Authors are grateful to them.

7. Secretarial assistance provided by S/Shri V N Shubham and Deepak Kumar of MODTECH, Delhi is gratefully acknowledged.

8. Last but not the least, authors thank God and respective families for providing ambient working atmosphere during the Covid period and thereafter for completion of this book.

CONTENTS

BUILT
ENVIRONMENT
SECTOR

HOUSING PROJECT AT FARIDABAD

During early sixties, Central Government was planning to shift a number of Government offices outside Delhi. As a part of this policy, the housing project was taken up at Neighbourhood-4, Faridabad (Haryana). This project was planned for construction of 2800 dwelling units of different categories from Type 1 to type 6 and six office blocks. Other buildings included Higher Secondary, Primary Schools, Community Centre. Shopping Centre etc. Thus, a mini township was to be developed at Neighbourhood-4, Faridabad. The development works included leveling and dressing, road work, footpaths, water supply, drainage, sewerage and external electrification. Besides, oxidation pond for cleaning of sewage, electric substations, horticulture development and tree plantation were also included in the total development plan. In first phase, fifty percent houses (1400 numbers) and three office blocks were constructed. All other buildings and total infrastructure for whole area was also sanctioned. One Circle Office headed by Superintending Engineer, three Civil and one Electrical Divisions headed by Executive Engineers were opened for this project. These offices were located at Faridabad. Most of the residents of adjoining neighborhoods of Faridabad were refugees from North West Frontier. A number of them worked as class II, III, and IV contractors for this project. They were straight forward persons and used to communicate without mixing of words.

By the end of the year 1965 most of the buildings were ready for occupation. Some offices from Delhi as also a unit of Indian Air Force were shifted to office complex of NH-4. For allotment of residences and office spaces, office of Assistant Estate Manager was opened at Faridabad, under Director of Estates, Ministry of Works and Housing, Govt. of India. After completion of major portion of project, the number of units of CPWD were gradually closed and only Faridabad Central Division was left. The Circle Office headed by Superintending Engineer did continue and some divisions from other areas were attached to this Circle Office. In December 1966, KB Rajoria, on promotion took-over as Executive Engineer, Faridabad Central Division. Most of the works under different contracts were completed but final bills for a number of works were yet to be paid. Besides, some works remained incomplete as contractors were not attending. Meanwhile maintenance of Govt. of India Press and its residential buildings were also attached to Faridabad Central Division. At that time, Shri SK Bose took over as Superintending Engineer, Faridabad Central Circle Shri DN Endlow was Additional Chief Engineer II, controlling Faridabad Central Circle. Shri NG Diwan Chief Engineer was head of CPWD.

When KB Rajoria took over charge as Executive Engineer, Faridabad Central Division practically had very little work load as most of the works were completed and about 50 final bills were pending. Besides, some works were left incomplete by contractors. Audit reports with hundreds of paras of all the three closed Faridabad Divisions were also pending. There were thirty Arbitration cases and five Court Cases. The task for clearance of arrears was undertaken under the guidance of Shri SK Bose and innovative solutions were found, for expeditious disposal.

Position of pending final bills was reviewed. In number of cases measurements were not fully recorded and Extra/Substituted items not even initiated. The responsibility for doing needful was given to newly posted Junior Engineers and Assistant Engineers, who were not familiar with these works. By persuasion and on giving assurance to them that they would not be questioned for mistakes of earlier incumbents, they all worked hard and finalized measurements. For a number of works which were completed earlier the test check was done by KB Rajoria. Besides, Extra/Substituted items were initiated by Junior Engineers and Assistant Engineers and in turn sanctioned, in Division Office or Circles office. Moreover, Reduced Rate Items were also initiated and sent to Circle Office for approval. Because of SE's personal intervention these were approved without delay. All the final bills were cleared in about two years period. Contractors could get their legitimate dues.

The status of incomplete contract works discussed by KB Rajoria with the SE at length and possible alternative solutions for completion of these works were examined. He dealt each case on humanitarian ground. Lot of time had already passed, on account of the fact that neither department took initiative to ensure completion of works nor contractors made any efforts. Inspite of the fact that lot of time had passed, full opportunity was given to contractors to complete their remaining work. Besides, department gave all help to contractors to ensure completion of work so long as they were serious. With this friendly approach, a number of contract works were completed. A few examples are given.

(i) **Construction of type four quarters:** This was a major contract and given to Shri Ramlal Hans contractor. The work was at stand-still for a few years. A notice was served on Shri Hans, with request to either complete the work in reasonable time or face termination. Shri Hans came to meet Executive Engineer and explained that he did not have enough money for completing the work but he had sincere intention. He requested that if some money could be given to him, the work could be restarted and he would complete the work. He had good track record of past, but was helpless on account of financial crises. The matter was discussed with Shri Bose and he gave a solution. To give financial assistance to contractor, for buildings which were completed more than six months back, part Security Deposit could be released. Accordingly, Security deposit was released and contractor gave an undertaking that he would utilize all the money for the project. The Contractor restarted the work. He was paid for the work done every fifteen days. With sincere effort of contractor and proactive attitude of department the work was completed.

(ii) The work of storm water drain was left incomplete by Shri MS Kalsi Contractor. A notice for termination was served. The contractor was given reasonable time to complete. He did not act. Thereafter, notice was given to him intimating that remaining work would be undertaken at his risk and cost. He came for meeting and requested for joint inspection. During inspection with him, it was noted that there were hindrances and at places site was not available, due to which he was not in position to complete remaining work. It was decided to close the contract.

There were thirty Arbitration Cases and in consultation with SE, it was decided to expedite action by department. The normal practice was that reply for Arbitration claims was drafted by Accounts Clerk, checked by Accountant and submitted by EE to SE for approval. In Circle office, Surveyor of works (SW) would examine the case. Thereafter, it was submitted to SE for approval and sent back to EE for further necessary action. When one such case was submitted to SE office, Shri Bose examined the case personally and called EE for discussion. It was a poorly drafted reply. He was kind enough to correct during meeting. Some portions of reply were dictated by him. He gave approved reply then and there. In view of this initiative by SE, all future cases were drafted by EE personally and directly submitted to him. These were corrected by SE in presence of EE and approved. The job was well done and well learnt. Lot of time was saved in disposal of cases and in-turn Arbitration proceedings were expedited.

There were five court cases in Tis Hazari Court Delhi. With effective persuasion of these cases there was expeditious disposal. Three cases were of Shri Navneet Ram Contractor. Arbitration Awards were against him and money was to be recovered. In the court, decree was obtained against the contractor but recovery was not possible as he declared himself as bankrupt. Satisfaction was that whatever could be done was done.

The Accounts of Faridabad Division were under the control by Accountant General, Punjab and Haryana, Chandigarh. There were 450 paras in audit reports and number of these were being processed as draft audit paras. With every inspection of Accounts Officer of AG, number of paras were increasing. A solution was found by SE, in consultation with the Accountant General. It was decided that an Audit team would be sent by Accountant General for discussion and settling of paras, if proper reply could be given. With mutual respect and intention to settle paras of audit reports, most of the paras were dropped. After all, efficiency of audit team was to be judged on number paras they would drop.

Unfortunately 75% newly constructed houses were having leakage from toilets and bath rooms. Therefore, dampness was seen in practically all the blocks. Residents were unhappy and they were put to inconvenience because maintenance workmen at Enquiry Office, could not satisfactorily attend to leakage. It was decided by Shri Bose that repair for attending to leakage should be done under a special drive for the whole colony. For this task addition plumbers and masons were employed. The job was well done and appreciated by residents. Besides, in most of the newly constructed houses, the paint on doors and windows was fading away. On detailed examination, it was noted that the wood used was not seasoned and paint used was not of appropriate quality. In consultation with SE, it was decided that all paint work should be redone. To make proper surface, all the old paint was burnt by gas blowers. Thereafter, primer and two coat painting was done. The paint was purchased by contractors from manufactures. The purchase was verified by department from manufacturers. Besides, manufacturer's representative was requested to inspect and verify that paint supplied by them was properly used. Thus, excellent painting work was done for all the houses and residents were very happy.

With Shri SK Bose, during beginning of carrier, it was real practical training to KB Rajoria and he was able to work properly and efficiently throughout service in the Government. Another person having deep knowledge about office working was Shri Muni Lal Sharma, Office Superintendent in Circle office. He helped in learning office procedures. The team work spirit was shown by other Engineers and office staff, which included Shri Mittal, Shri Bindal, Shri Satija and Shri Madan Assistant Engineers as also Shri Kasturi Lal and Shri Hora Junior Engineers. A very special experience of Faridabad was dealing with local contractors who came from North West Frontier Province area and settled in NIT Faridabad. They were honest in their actions and were always willing to work sincerely without mixing of words. Names of Shri Hem Raj and Shri Sher Singh are worth mentioning.

LESSONS LEARNT

(i) **Contract Management:** A number of contracts were awarded for the project. Except for two contracts, all others were executed in proper manner and work was completed. Out of two contracts one for a building project, the contractor was willing to complete the work, but did not have funds. For this project a solution was found within framework of Government rules and regulations. This enabled contractor to complete the work. Meeting cash flow requirements by timely payments to contracting agency is an important element of successful completion of a work. Thus, efforts to find a solution to the problem and then implementit is a good approach for contract management. For another contract, depending on specific situation, that was non availability of site, closure of contract was allowed. The contractor was not at fault and proper action was taken.

(ii) **Finalizing accounts, bills, and arbitration cases:** A project essentially comprises pre construction, construction and post construction stages. Usually, the major part of activities of first stage are carried out in planning office, the remaining two are essentially carried out in the field units. Concluding or *samapan* stage of a project is as important as a good start. It is generally seen that Government agencies are put to difficult position, when after completion of projects, field units are closed or shifted. Result is that final bills, audit paras and arbitration cases are not given the required importance. For this project, a seasoned senior middle level engineer, knew the job and took initiative for finalizing accounts, bills and arbitration cases. Engineers must take such initiative, in the overall interest of the Government.

(iii) **Maintenance:** Buildings are required to be maintained properly, for safe and comfortable living of allottees. Besides, for durability and useful long life of buildings, proper maintenance is necessary. Any deficiency in workmanship or specifications during construction stage requires rectification during maintenance. In this project, it is worth mentioning that the deficiency was analyzed objectively and attended in a systematic manner during maintenance stage. This provided great satisfaction to the users, besides preserving and enhancing life of the buildings. A good decision making is critical for maintenance.

(iv) **On-the-job Training and mentoring:** Direct recruit class 1 officers get a post of responsibility at middle level like that of Executive Engineer, after training and a short posting at Junior Level. To work properly at middle level, guidance of senior level officers is required. In this project, this job of giving practical training and ensuring efficient working of Executive Engineer, the Superintending Engineer gave guidance. For efficiency of an organization this type of initiative by seasoned officers is considered necessary and desirable.

CAPITAL OF ARUNACHAL PRADESH

Arunachal Pradesh was earlier known as NEFA (North East Frontier Agency). During 1971, it was declared as Union Territory, headed by Lt. Governor. At that time, headquarter of Arunachal Pradesh was located at Shillong, Capital of Meghalaya. Shri KAA Raja took over as first Lt. Governor of Arunachal Pradesh. During that period Central PWD was main construction wing of the Government. Several CPWD Divisions were located at different places. These were entrusted with construction and maintenance of buildings, roads, micro hydel projects etc. Three Superintending Engineers and one Senior Architect were working under Arunachal Administration and their headquarters were also at Shillong.

It was an administrative necessity to locate capital of Arunachal Pradesh within geographical area of the Union Territory. During 1973, it was proposed to locate capital at a place near Naharlagun village in Subansiri District. Hon'ble Lt. Governor directed Superintending Engineers and Senior Architect to visit the area and report about suitability. Accordingly, Superintending Engineers, S/S S Nagarajan, RA Khemani and KK Reddy and Senior Architect Shri RK Agarwal with other Senior Engineers visited the area. The capital site was approachable from Doimukh, located in foothills of Subansiri District, and connected by road with the National Highway in Assam, on the north side of river Brahmaputra. All team members met at Doimukh and after crossing river Dikrong by boat, went on foot to Naharlagun village (about 9 km), mostly along the bank of river. It was noted that even at Naharlagun village some land was available. Thereafter team members went uphill towards proposed permanent capital. It was found that a flat terrace was available, which was considered suitable for relocating the Capital. In fact, additional land was also available across the river. At this location there were remains of an old fort made of bricks. After the visit, a report recommending the suitability of the site for capital was submitted. The report was accepted and the place selected for capital was given a new name 'Itanagar'. The new name of the capital had taken inspiration from fort remains made of bricks.

Hon'ble Lt. Governor was desirous for shifting of his office within Arunachal Pradesh at an early date. He visited the area and decided that a temporary capital should be located at Naharlagun. Hon'ble Lt. Governor directed that all three Superintending Engineers should work jointly to ensure that work of temporary capital is completed in about one and half year time. To facilitate working, two new Divisions were sanctioned by the Government for timely completion of the work of the temporary capital. At that stage, Shri RK Sundaram took over as Superintending Engineer one and Shri BV Subramanyam as Superintending Engineer two. Shri KK Reddy continued in his place.

To implement the project of temporary capital expeditiously, staff was posted for newly sanctioned units and distribution of units under three circles was decided. Capital Division One was placed under Circle 1 and Shri BK Biswas joined as Executive Engineer. Capital Division Two was placed under Circle 2 and Shri HKL Mehta joined as Executive Engineer. Three sub divisions were taken from these two Divisions and placed under Pasighat Division where K B Rajoria was Executive Engineer (under Circle 3). Thereafter three Superintending Engineers and Senior Architect met at Doimukh. Executive Engineers also joined the meeting. It was decided that the area should be made approachable by constructing a Pontoon bridge on river Dikrong near Doimukh and excavation along hill side should be done to make a jeepable road upto Naharlagun. Besides, jungle cutting and tree cutting should be done to ascertain extent of land available for building of temporary capital. It was a forest with no human settlements nearby. Trees were very tall with the girth size of 3 to 4 meters. Surprisingly the roots were not very deep. The number of snakes was innumerous on the bushes around and it was really scary for the staff. However, with the help of local tribal men engaged as daily labor, work of cutting the bushes and trees was accomplished. It was day and night work of Shri HKL Mehta and his team that jeepable road (about 10 km) was made in about two months time. The road work was highly appreciated. Meanwhile, the jungle clearance work was also completed and thereafter field survey was carried out to make a layout plan for temporary capital site.

The Senior Architect, in consultation with Superintending Engineers prepared layout plan of temporary capital. The identification of priority was very meticulously done by the Senor Architect with the help of Engineers. These plans were submitted for approval of Hon'ble Lt. Governor, and thereafter prepared preliminary plans of buildings to be constructed on layout of the area assigned in the master plan. Non residential Buildings to be constructed were:(i) Secretariat Building, (ii) Arunachal Bhavan (Circuit House), (iii) High School Building, (iv) Post office, (v) Fire Station and, (vi) Hospital Building. Besides it was decided to construct 550 residential quarters from type I to type IV categories. External services developed were: (i) Black topped approach road with side drains and culverts from Doimukh to Naharlagun, (ii) Water supply mains from water source at uphill, sedimentation tanks and sump, (iii) Electrical Power Generators, (iv) Water supply distribution lines, (v) Internal roads, vi) storm water drains and outfall drain to nearby existing nallah as also (vii) Electrical distribution network.

For deciding specifications of buildings, there were several meetings at Doimukh, attended by Superintending Engineers, Senior Architect and Executive Engineers. Use of local materials, skill and technique was incorporated to ensure success and sustainability. It was decided to have single storied buildings with isolated RCC footing and pedestal, wooden columns with wooden roof truss. On walls Bamboo ekra (woven bamboo strip) with cement plaster and C.G.I. roofing was decided. Door and windows for building made of local wood were provided with glazing. Electrical wiring was provided on surfaces. The specifications decided were based on the climatic conditions and prevalent practices in the area so that it could be implemented with the semi-skilled workers available in the area. After finalizing specifications, working drawings and detailed drawings were prepared by the Senior Architect. The Preliminary Estimates were submitted to the Government for approval and these were approved on top priority. In order to facilitate working at temporary capital site, local administration deputed adequate police personal for patrolling of the area. Besides, Forest department deputed guards to ensure that elephants and other stray animals remain away from construction site. Moreover, local administration issued blanket permission for CPWD to cut trees required for construction of the roads and buildings. Supply of food and vegetable was also arranged by local administration. All required statuary and logistic facilitation were extended to CPWD for implementing the work.

The preliminary work to start the project was initiated at project site. For taking quick decisions several meetings were held at Doimukh and later on the venue shifted to Naharlagun. The command of the project was jointly held by three Superintending Engineers. Hon'ble Lt. Governor desired that Superintending Engineers should stay at Doimukh/Naharlagun, for ten days each by rotation. Besides, they should monitor the work and give guidance to all the engineers irrespective of jurisdiction. It was a unique administrative arrangement. There were changes of incumbency during implementation period. Shri RK Sundaram took over from Shri S Nagrajan and Shri BV Subramanian from Shri RA Khemani. The control of project remained equally effective, even after changes. During meetings at site, decisions were taken regarding manner of implementation of the project and the work was monitored. The normal procedure by call of tenders for full or part work was ruled out. This procedure would have taken lot of time. Besides, even Class II contractors, were not available in the area. Moreover, even for smaller size of packages, the award of work on call of tenders was not found practicable. Therefore, it was decided that work would be awarded on piece rate or labour rate to contractors whoever was desirous to work. The size of package was decided on the basis of contractor's capacity and capability. It was also decided that if the contractor was unable to perform, balance work would be taken out of his hands or reduced in size, without any written notices. It was also decided that as for as possible materials would be procured on Rate Contract. Thus, materials like steel, cement, glass, sanitary ware, paint, CGI Sheets etc., were procured on Rate Contract. Timber was also required in huge quantity. A meeting was called, in which all the timber suppliers of the vicinity were invited. After negotiations, a reasonable rate for supply of timber was agreed by them. For other materials, like fittings and fixtures, Executive Engineers were authorized to purchase from market of nearby places. The direction was very clear that everybody should work day and night. They should ensure that work should not be stopped.

The work of developing services proceeded simultaneously. For water supply, a storage tank was constructed, uphill and water brought to this tank by gravity from a nearby river. The water supply mains were laid along the roads, and connection given to individual units. Similarly, generators were installed at suitable location and distribution lines laid to give connection to individual units. Road and drains were also developed as per the plans, to have proper connectivity. For disposal of sewage, septic tanks were provided. All the services work was done simultaneously with building works. With mobilization of artisans by contractors, and making arrangement for materials directly by department, the work proceeded at very fast pace. In fact, whole area was looking like a big fair where everybody was busy. Besides, there was feeling of competition amongst various Assistant Engineers and Junior Engineers. To facilitate working during night, temporary lighting of the area was arranged. Initially there was anxiety. As work proceeded the confidence developed. All Engineers, contractors and artisans had only one desire and wish that work should be completed as early as possible. With sincere efforts of everybody and under the command of three Superintending Engineers, the project was completed in about one and half year. It looked like a miracle.

Hon'ble Lt. Governor inspected the project and took round of whole area. Superintending Engineers accompanied him during the visit. He highly appreciated efforts of CPWD Engineers. Recognition of work by political and administrative leadership had boosted the morale of CPWD Engineers. Immediately thereafter, the Arunachal Pradesh Government offices shifted to Naharlagun, the temporary Capital of Arunachal Pradesh. Examples of such fast work under difficult circumstances are very rare. All Engineers who worked for this project were very happy. All the Superintending Engineers viz S/Shri S Nagrajan, RA Khemani, RK Sundaram, BV Subramanian, KK Reddy as also Shri HKL Mehta Executive Engineer were awarded Gold Medals. Silver medals and commendation certificates were awarded to other Engineers. Shri KB Khanna and Shri Rao Assistant Engineers of Pasighat Central Division were awarded Silver Medals. After the Arunachal Pradesh Government shifted to temporary capital, the work for permanent capital started.

LESSONS LEARNT

(i) **Leadership** - Great leaders fix tough targets and ensure that target is achieved. They encourage and provide all the help and direction. For this project, it was the leadership of Col. KAA Raja (retired), Hon'ble Lt. Governor, that impossible was made possible. Delegation of powers by the Lt. Governor was the key to success. In normal course the project would have taken about four years. Thus, it is clear that if there is a will, impossible can be achieved. For Engineers, it was a great challenge and command was jointly held by three senior level officers. It was made possible by hard work, foresight, and administrative skill of these officers, although two officers were posted out as part of routine transfers during the project implementation period and new officers joined in their place. It is worth to mention that it speaks high of Central PWD and culture of organization.

(ii) **Clarity of targets and micro-level planning** - What was to be done was clearly decided in the beginning itself. The layout, the type of buildings, the structural frame and the finishing were decided and approved. There was no scope for keeping any decisions pending. Modes of procurement of materials, agencies for implementation were also decided so that no time was lost in ensuring their availability. Advance micro planning paid dividend and decision making was brought to project door steps. Of course, the driving force was the direction from the Hon'ble Lt. Governor that the work should not stop due to any reason whatsoever. Thus, senior and middle level officers were essentially required to ensure that there was no shortage of required resources. Even the external factors like protection from wild animals, arrangement for food supplies, law and order conditions was planned and built into the working arrangements.

(iii) **Professional commitment and hard work** - All Executive Engineers, Assistant Engineers and Junior Engineers worked almost round the clock, even when living conditions were difficult. It was their dedication and sincere hard work that targets could be achieved. Contractors, masons, carpenters, electricians, plumbers and other workers lived under difficult conditions. They worked day and night. They were motivated and were paid timely. They deserve appreciation. Besides, it was the backup of administrative officers, Food and Supply Department, Police, Forest Department as also other wings of Government and their commitment to make living conditions safe and comfortable, which enabled engineers to work. They all deserve appreciation.

(iv) **Stakeholder's Delight** - Completion of this project well within the time frame decided by the Administration was highly appreciated by the Government, district administration, local leaders and public. The system of acknowledging services by hardworking engineers by rewards paid its dividend in this prestigious project as well. The credit was given to the concerned sincere senior, middle as well as field level engineers.

Pasighat was the oldest township of Arunachal Pradesh. It was founded in 1911 AD by the British as a gateway to the Abor Hills and the North East area in general. Pasighat has always remained an important place of Arunachal Pradesh. This township is situated on right bank of river Brahmaputra. The river Siang, the main tributary of river Brahmaputra enters plains near Pasighat and joins with river Lohit, another tributary and becomes river Brahmaputra. Pasighat Central Division of CPWD was established during mid fifties and development of township as also construction of roads was undertaken by this Division. By the end of sixties, it was a beautiful township with residential buildings, hospital, schools, college, agriculture institute, radio station and all other departments of Government under the administrative Control of Additional Deputy Commissioner. Agriculture, horticulture, and tourism continued to be the main source of economy for the town. All buildings and their services were maintained by Central PWD. In January 1971, KB Rajoria took over as Executive Engineer at Pasighat. There was also an Electrical Subdivision at Pasighat under Jorhat Electrical Division. Several building and road projects were undertaken at Pasighat, out of which three important projects at Pasighat are described here.

JAWAHARLAL NEHRU COLLEGE

Jawaharlal Nehru College at Pasighat was established in the year 1964. It was the first institute of Arunachal Pradesh for graduate studies. It was expected that students of this collage would play an important role for development of Arunachal Pradesh. The land for collage complex was on the north side of Pasighat township. Ordinary Basha (OB) type buildings were constructed to run the collage. These were temporary hutments made of woven bamboo strip walling with thatch or CGI sheet roofing and having only a few years life. A permanent building complex was necessary in order to run the collage in a proper manner. Earlier some initiative was taken but the college building could not be constructed as the resourceful contractors were not available. The status was reviewed and issue regarding award of work of college building to a contractor was examined. Engineers were of the opinion that it would take about two years to complete the project if a contractor could be found. But such contractor was not available. During March 1971, Shri Krishnatri, Development Commissioner, came to Pasighat and, visited the college complex. The principal, faculty members and students of the collage requested Shri Krishnatri that the permanent collage building should be constructed on priority, because the proper teaching and learning was not possible in temporary hutments. In turn the Development Commissioner discussed with Engineers and desired that the collage complex should be completed before start of next academic session. It looked almost impossible and before giving him any commitment, Executive Engineer requested him to allow internal discussions, to find ways and means to complete the project. After brainstorming session and review of the situation, Engineers came to conclusion that incase tendering was again done to award work to a contractor, there would be uncertainty. It was difficult to find a resourceful contractor. The other alternative was to undertake the work departmentally. But there were difficulties. Engineers neither had financial powers to purchase materials nor fix labour rate contracts without call of tenders. Besides, there was uncertainty regarding availability of building materials at Pasighat, Dibrugarh and Guwahati. Only at Kolkata, all the materials were available at reasonable rate.

Different alternatives were discussed at length with Development Commissioner. It was proposed to him that the building could be constructed expeditiously before the next session only if Government gave special approvals. It was submitted that building materials should be purchased from Kolkata on the basis of 'on the spot' quotations. For these purchases Engineers would travel by air to Kolkata. For employment of labour contractor permission was to be given to award job work to petty contractors and without call of quotation/tenders at reasonable rates. As desired by him, a proposal was given in writing for approval of the Administration. The proposal was discussed at length at Shillong. Shri KK Reddy, Superintending Engineer, supported the proposal as he was confident about capability of field engineers to implement project. It was approved by the Administration. After receiving the approval, without taking time KB Rajoria along with Shri Ghosh Junior Engineer proceeded by air to Kolkata. There was no difficulty for making payment as cheque book of Reserve Bank of India was available and CPWD was a known entity. Immediately after reaching Kolkata, the team contacted shopkeepers, wholesalers and manufacturers, inspected building materials and selected. Thereafter, on the spot quotations were collected. After comparative study of quotations, Supply Orders was placed at the at lowest rates. In fact, it was found that market rates at Kolkata were less than half the rates at Guwahati or Dibrugarh. Immediately materials were collected, payments made to suppliers and materials sent to the godown of cartage contractor for transporting to Dibrugarh. After reaching back to Pasighat, petty contractors were selected for different jobs. Meanwhile, all materials reached Dibrugarh. Engineers went there and brought materials by boat. In order to work day and night, the whole college complex was lighted with flood lights. Shri Dalip Singh Assistant Engineer and Shri RB Chaudhary Junior Engineer did outstanding work for this project. They worked very hard with all worker and supervisors. Roofing work was done on first priority, so that other jobs could be done even when it would rain. Thus, work continued during heavy rains even though the average rainfall at Pasighat was more than 200 inches (500 cm) annually. As the work proceeded, it was felt that impossible was becoming possible. This work was could not be undertaken for a number of years but with change of strategy, could be completed now in a few months' time only. The intimation regarding completion was given to authorities at Shillong. This achievement was highly appreciated. The building complex was inaugurated by

Shri Govind Narain, Secretary to Govt. of India. He was impressed with the work done by Central PWD Engineers. The Principal, faculty members of the collage and students were very happy with the completion of the college building.

MICRO HYDEL PROJECT

During 1971, the electricity was supplied to the township by Diesel Generators. The responsibility of generation and distribution was that of CPWD. The electricity was supplied during evening hours only.

Arunachal Pradesh had huge potential for hydel power generation. But it was not exploited. In Arunachal Pradesh the responsibility for identification, survey and detailed planning of Micro hydel projects, was given to Central Water and Power Commission (CWPC) Govt. of India. They identified a few projects including one near Pasighat. The Micro hydel project was to be developed on a small tributary to river Siang, a few kilometers away from Pasighat. Detailed drawings for this project from Central Water and Power Commission were received. After financial approval from Arunachal Pradesh Administration, the responsibility was given to CPWD to develop and execute this project. CPWD officers did not have any experience for Micro hydel projects. Therefore, it was unique opportunity as well as a challenge.

Important components of this Micro hydel project were: Weir across the river, Collection Chamber adjacent to the Weir, Pipeline from Collection Chamber to the Storage Tank, Storage tank, Penstock pipe, Power house building, electricity Supply and Overhead lines. After detailed discussions with Shri KK Reddy, Superintending Engineer, it was decided to execute the project departmentally. Based on experience with the College building, it was decided to depend on Kolkata market for purchase of different components, and not on local market of Assam. Accordingly, quotations were called for purchase of different components of the Project. This included fabricated grills for the weir, gates for collection chamber, valves, penstock pipes, turbines, overhead lines, electric cables and poles for supply line etc. Fortunately, good response was received from Kolkata suppliers. After receipt of quotations, proposal was submitted to Arunachal Administration at Shillong. Shri Reddy ensured quick approvals.

Immediately thereafter, the work was started departmentally. The Weir structure was developed in two parts by diverting the flowing water of the river. There was no difficulty in constructing the Collection Chamber – a RCC tank adjacent to the river, because rock was found. It was constructed about half km away. The pipe line from Collection Chamber to the Storage tank was laid on valley side of the road. The construction of RCC Storage tank, was started after a level ground was developed by excavating the hill having gentle slope. From the Storage tank, electric resistance welded (ERW) Pipes were laid up to turbine room at lower level. It was the most difficult task. After turbines arrived, it was expected that the installation etc. would be completed in about 45 days period and project could be commissioned. The date of completion was accordingly intimated to the Superintending Engineer.

An immediate reply was received from Superintending Engineer that the Lt. Governor would inaugurate the project after thirty days. The team was put in a difficult position. In consultation with engineers and contractors, it was decided to organize day and night shifts. For lighting the area, Generators were installed and electrical lines were laid. All facilities were given to workers for comfortable working and Engineers visited the project site in shifts. The work progressed well and it looked that target could be achieved. About one week before the schedule date of arrival of Lt. Governor, it was informed that he preponed the visit by one day. It was something unexpected and Engineers were nervous. It was not like completion of a building project. In this project, electricity was to be produced through turbines. There was no way except to work for 24 hours a day. Even one day before date of inauguration, there were doubts about generation of power. With focused attention and hard work, turbine started working and produced electricity from the turbines, only a few hours before arrival of the Lt. Governor.

Lt. Governor along with Superintending Engineer reached the project site, as scheduled. The public of Pasighat and adjoining villages gathered in good number to welcome him and participate in the inaugural function. It was a superb programme of inauguration of completion of a difficult and tedious project and of high appreciation by the Lt. Governor.

INGLISH BRIDGE

While inspecting the Store Complex of Division office, it was found by the Executive Engineer that a number of pipes with special joining links were lying in the Store. Enquiries revealed that these pipes were lying in the store for a number of years. On further investigations, it was informed that these pipes belonged to British Army and lying since Second World War era. Besides, local enquiries revealed that these pipes were meant for fabrication of a bridge, known as Inglish Bridge. Pipes and joints were in good condition and could be used for construction of a single span bridge of 240 feet (about 72 m).

At that time Pasighat - Jonai Road was the only link of Pasighat to the rest of the country. Jonai was a small town in Assam near a railway station known as Murkang Selek. The road from Jonai to Pasighat (40 Km), was constructed and maintained by CPWD. There were number of timber bridges on this road. But at one location that is on Leku River, the span was about 240 feet. The span being large, the timber bridge used to become vulnerable during heavy rains. Sometimes there was breach. It was decided to construct a permanent bridge at this location with Inglish Bridge parts. Bridge components were inspected in detail for the purpose of fabrication of bridge and found that some more components were required. On enquiry, Shri Tarung Naung, a local contractor, mentioned that it could be possible to find missing parts with scrap dealers at Dibrugarh. He was sent to Dibrugarh and he could find missing parts through his diligence. The work was awarded to Shri Tarung Naung and he procured missing parts. Masonry abutments were constructed. Stone and sand were abundantly available nearby so there was no difficulty for procurement. Work of centering and shuttering was done departmentally by procurement of timber. CPWD fabricated and installed the bridge in about two- month period under the leadership of Shri AK Sarin, Assistant Executive Engineer. Thereafter bridge work was completed. Construction of this bridge helped to have proper permanent road connectivity with Jonai and in turn Murkang Selek Railway station, the life line of Pasighat.

GOVERNMENT'S DELIGHT

Construction projects implemented at Pasighat were highly appreciated by Arunachal Administration. KB Rajoria, Executive Engineer was awarded Gold Medal, Shri AK Sarin, Assistant Executive Engineer and Shri Dalip Singh Assistant Engineer were awarded Silver Medals. Shri RB Chawdhry Junior Engineer was awarded Commendation Certificate.

LESSONS LEARNT

(i) **Project Risk and Mitigation:** CPWD had prescribed rules and regulations as also methodology to be followed for implementation of projects. At a place like Pasighat, situated in deep interiors, a small project like construction of college building could not be executed for a number of years, by following the normal procedures. When the challenge for completion of the project was given to Engineers, they had brain storming session and came out that procedural issues beyond their administrative and financial powers required approval by Administration. This proposal was accepted and the project was implemented in a few months' time. Thus, for proper implementation of a project, Engineers should always analyse the specific requirement and get approvals from competent authority. It is an example of implementation under difficult circumstances at a place where required resources were not available.

(ii) **Goal Setting and team work:** Building up and developing confidence among team members for success of any project is very important. With the brain storming session with the team, it could be decided as to how the deadlock could be broken. Thus, spending time to locate a resourceful contractor and then be at his mercy was not acceptable. The team wanted to achieve the target of completing the building before start of the academic session and found ways and means of doing so by taking up the onus of taking all the responsibility of procurement of materials, transportation, arranging labour and supervise the implementation on their own shoulders. Certain approvals from the Government of Arunachal Pradesh were required which the Administration having confidence, readily gave.

(iii) **Professional Commitment and resource Management:** The micro hydel project could also be implemented by arranging materials from the place where availability at economic cost was assured. The experience of similar exercise in case of college building was made use of here. Besides, there was will power and commitment for hard work and motivation, that the project was completed even at preponed schedule. Breaking up of the project into different components and then arranging resources and taking up the construction work ensured full control over the activities. Having arranged required resources, increase in number of working hours was the only way out to achieve the preponed target set out by the Lt Governor. Thus, detailing of activities and providing matching resources for the same was essential for success.

(iv) The construction of Inglish bridge is an example where available resources could be utilized at proper place. A very useful resource in the form of components of a bridge was lying idle for nearly three decades before these were noticed. It was innovative thinking which made this project possible. The important lesson is to make use of all the available resources and convert them into asset.

In all the above projects one could see that the CPWD team had tremendous desire to do things and made all possible efforts to do it. The 'mool mantra' was that 'I want to do it and I shall do it'.

(v) **Stakeholder's delight:** The Arunachal Administration, on their part, were eager to bring the fruits of development to the area at the fastest pace. That is why, after ascertaining the potential of the delivery system, did not hesitate to set out tight targets. The system of awarding the hard work through medals or commendation certificates kept the morale of the CPWD team very high. It encouraged one and all to do their best. Thus, acknowledging as well as rewarding always helps in smooth and efficient functioning in any organisation.

4

SANSDIYA SOUNDH (PARLIAMENT SECRETARIAT EXTENSION BUILDING) AT DELHI

After Independence of the country, activities at Parliament House Complex gradually increased. Offices of Secretary General Lok Sabha and Secretary General Rajya Sabha were located in Parliament House. As the staff increased, there was not enough space for the staff and difficulties were felt. Besides, for national and international conferences, proper and adequate conference halls were not available. During early seventies, it was decided by Government of India to construct a multistoried building, on the plot adjacent to Parliament House. In this building conference halls of different capacities were to be developed, along with other facilities for organizing conferences. Besides, office space to accommodate Parliament Secretariat was also to be provided. Central PWD was assigned this prestigious building project. The plans were prepared by Shri JM Benjamin, Chief Architect, and finalized in consultation with Parliament Secretariat Authorities. According to these plans, lower ground floor, ground floor and first floor were to provide for conference and VVIP chambers. Remaining five floors were for Lok Sabha and Rajya Sabha Secretariat. Important spaces provided in ground and first floor were as follows:(i) Ground floor- (a) Entrance lobby (b) Ornamental Staircase upto first floor, (c) Large Committee Room for 150 persons seating capacity and (d) Wide corridors for exhibitions etc. (e) Banquet Hall (f) Auditorium (g) Four small Committee Rooms for 50 persons each.(ii)   First floor- (a) Three office Chambers for VVIPs and (b) Two office chambers for Secretary Generals.

The Preliminary Estimate for this project was prepared in the office of Chief Engineer, Zone 1, CPWD and submitted to Parliament Secretariat for approval. After Administrative Approval and Expenditure Sanctions of the project, the detailed drawings for the building work were drafted by office of Chief Architect. Meanwhile, the approval was obtained from New Delhi Municipal Council. This project was assigned to Construction Division Seven under Delhi Central Circle Four. The building work, water supply and sanitary installation work, internal and external electrification work, air conditioning work and other works were awarded to different contractors and by December 1974, most of the works were completed.

During 1974, Parliament Secretariat gave sanction for interior finishing work, fixtures, furnishing work etc, for the ground and first floor. This work was to be done before Commonwealth Conference, scheduled in 1975. For planning and designing of interior decoration, furniture and finishing works, it was decided to engage services of private interior designers and architects. The work was awarded to two agencies, viz M/s Sachdeva Eglistan and M/s Rajendra Kumar and Bose Associates. It was decided by the Secretary General, Parliament Secretariat, that they would like to approve interior finishing and furniture work. For undertaking furniture and finishing work, some specialized agencies were shortlisted, in consultation with private designers. The limited tendering was done so that only competent agencies get the work. Even otherwise most of the interior decoration work was not covered by CPWD Specifications. The CPWD was directed to expedite the work and complete before the date fixed for the conference. In May 1975, K B Rajoria took over as Executive Engineer from Shri CB Lal. At that time Shri RJ Bhakru was Superintending Engineer (Civil) and Shri GK Khemani was Superintending Engineer (Electrical). Shri Dasgupta was Executive Engineer (Electrical). Shri KR Jani was Architect, Shri Ashok Agarwal was Assistant Executive Engineer (Civil) and Shri KK Sharma was Assistant Executive Engineer (Electrical). Besides, Shri Anand and Shri Amarjeet Singh were Assistant Engineers (Civil). Within a short period all work related to finishing, services, internal electrification, fire-fighting equipment, veneering works on walls, flooring, horticulture work, kitchen equipment, furniture carpets etc was to be completed. It was a very difficult task and time left was not much.

When the position of progress of work was reviewed, it was noted that the furniture and finishing work was not proceeding at required pace, as approval from Secretary General was taking considerable time. There was a long drill. First the furniture was to be made under direction of interior decorator, then it was to be seen by field staff, thereafter approval by Chief Engineer and finally by Engineer-in-Chief CPWD and Secretary General jointly. Thus, considerable time was taking for approvals. In fact, Engineer-in-Chief (Head of CPWD) and Secretary General were very busy persons and finding suitable time was difficult.

During that period, Shri VR Vaish took over as Engineer-in-Chief CPWD. As per the practices, office of Secretary General contacted Shri Vaish for joint inspection, for approvals. Instead of accepting the request for joint inspection, Shri Vaish informed the Secretary General Office that for the inspection of furniture and finishing work only Chief Engineer would join because he was equally competent and experienced Engineer. Besides, responsibility for the project was fully with the Chief Engineer and he would give final decision on behalf of CPWD. Thus, Shri Vaish, ensured that working become comfortable for Shri OP Mittal Chief Engineer. He gave Shri Mittal full authority to deal with Secretary General.

Thereafter, when the project was jointly inspected by Secretary General and Shri OP Mittal Chief Engineer, it was decided that for timely completion of project, full responsibility and authority should be with CPWD Engineers. It was agreed that Secretary General would take up only very important issues and not participate for approvals of finishing and furniture work. This was the turning point. Shri OP Mittal Chief Engineer became final authority and decided that every morning, he would inspect the project, take required decisions and approve the work. Contractors were told squarely that they should complete the work, as per programme, otherwise balance work would be taken out of their hand. Shri JM Benjamin Chief Architect used to spend considerable time on the project. Shri KR Jani, who was assisting him, could appreciate urgency of project and conveyed decisions of Chief Architect without taking time.

For remaining interior work, furniture, equipment, furnishing etc., several short term tenders were invited from listed agencies. Most of jobs were unique, like veneering work on walls, carpet, kitchen equipment, decorative lights, auditorium chairs etc. All engineers, contractors and artisans worked day and night to ensure completion of project. It is worth mentioning that Shri Vaish Engineer-in-Chief did inspect the project on several occasions. He gave very useful decisions. On one occasion, when Shri Vaish was visiting the project, we were informed that Honorable Speaker would arrive on an unscheduled visit. The immediate reaction of Shri Vaish was that Shri Mittal and the team of engineers, should attend the inspection of Honorable Speaker. He himself decided to leave the building through an alternative route, so that meeting Honorable Speaker could be avoided. Shri Mittal with the team of Engineers met Honorable Speaker. Fortunately, Honorable Speaker was satisfied with the progress of work. The environment became favorable and CPWD continued to work round the clock.

As the work proceeded, Shri OP Mittal was happy that on account of sincere hard work by Engineers, and contractors the project was getting completed. Still there was anxiety on account of emergency in the country. One day, when Shri Mittal was on his routine morning inspection, we were informed that Honorable Prime Minister Indra Gandhi along with Honorable Speaker Shri Gurdayal Singh Dhillon, had already entered the complex, along with Secretary General Lok Sabha. In fact, no earlier information was given to CPWD for this visit. Honorable Prime Minister was inspecting the building and we were standing in one corner of the building with anxiety and uncertainty. It was a very tense situation. After Honorable Prime Minister reached office chamber meant for her, one senior officer from Lok Sabha Secretariat came and informed that Honorable Prime Minister desired to meet Engineers working for the project. We proceeded towards her office chamber. Anything could happen and we were prepared for the worst. When we met Honorable Prime Minister, she was very happy and expressed her satisfaction and appreciation for the work done by Engineers. We were mentally relived and were in turn very happy. It looked that all the hard work has brought good results. In fact, Shri OP Mittal was the happiest person. After a few days, we were informed that Honorable Prime Minister mentioned about this project to Honorable President of India, that she never saw such a beautiful building in the country. After a few days Honorable President of India came for inspections of the building and he was also highly impressed.

The project was timely completed for Commonwealth Conference and it was hosted in this building in befitting manner. The work done for this project was highly appreciated by all participants. After a few days, Shri OP Mittal Chief Engineer was honored with "Padma Shri" award for successfully completion of this project. All the engineers were given cash award and appreciation letters. It was a pleasant conclusion of a very difficult project.

LESSONS LEARNT

(i) **Improvement in Communication and decision making**: At times, on account of over anxiety to complete a project, the top level of client department starts coordination with the of top level of implementation agency. This project is an example of such initiatives by Parliament Secretariat. Disadvantage of this type of communication was seen by Shri VR Vaish, Engineer-in-Chief. He took initiative to stop such communication. The project could proceed well only after head of implementation unit, that is Chief Engineer, was given full authority.

(ii) **Leadership:** The team work spirit was put in the project by Shri OP Mittal Chief Engineer. After he became final authority, project could proceed at fast pace and completed in time. Having installed confidence among the team members, the leader and the members proceeded to perform their assigned jobs sincerely without being afraid of the ensuing outcome. They all believed in carrying out their 'karma' or duty instead of being worried about 'phala' or the result. Satisfactory completion of the project before due date was the biggest reward for the team.

(iii) **Client's delight:** Appreciation by top personality of the country was a memorable experience. Of course, Awards did give satisfaction of achievement.

Parliament Secretariat Extension Building -
Front Elevation from Parliament Street

Parliament Secretariat Extension Building -
Outside entrance area

Parliament Secretariat Extension Building -
View from Parliament street showing entrance lounge

Parliament Secretariat Extension Building -
Courtyard and lounge for conference rooms

5

RESETTLEMENT COLONIES OF DELHI

I. PLANNING AND EXECUTION

During 1973-74, Socio-economy survey was conducted by Town and Country Planning Organization, under Ministry of Urban Development, of all Jhuggi clusters of Delhi, to ascertain status of society in terms of physical, social and economic infrastructure. In all 1373 Jhuggi clusters were surveyed. It was noted that they had no water supply, street light, toilets and any other living facility. In these Jhuggi clusters more than 1,40,000 families were living.

In order to makes their living comfortable, it was decided by the Government of India to provide them alternative suitable place to live. For this purpose, sixteen Resettlement Colonies were planned by Delhi Development Authority (DDA), under 'sites and services' scheme, in geographical boundary of Delhi. The total area under this scheme was 968 hectares and 1,48,262 plots of 21sqm were developed. The area meant for facilities was 61% and for plotted development 39%. The basic concept was developed by planning 500 plots of 21sqm (3m × 7m) in each cluster. The distribution of area in each cluster was, 32% for plots, 18% for roads, 16% for parks/play grounds/ open spaces and 4% for commercial use. Space was kept reserved for Cinema Halls, Fire stations, Police stations, colleges, hospitals etc. Basic services were also conceptualized. For water supply one hand pump for 31 families and one water hydrant for 70 families was planned. For sanitation, common toilet blocks were planned by providing one toilet seat for seven families. Street light poles were planned for roads. Drain and outfall drains were planned for all the colonies to be developed. The colonies were conceptualized after giving due consideration to provisions of Master Plan. Total sixteen colonies were planned having 1,48,262 plots. Different colonies were,

  (i) Dakshinpuri & Extension

  (ii) Khanpur

 (iii) Chaukhandi

 (iv) KhayalaComplex

  (v) Gokulpuri

 (vi) Shakarpur Complex

 (vii) Nand Nagri

(viii) Sultanpuri

 (ix) Mangolpuri

  (x) Hyderpuri

 (xi) Jahangirpuri

(xii) Patparganj Complex (Khichripur, Kalyanpuri, Trilokpuri)

(xiii) New Seemapuri

(xiv) Nangloi

 (xv) N.G. Road and

(xvi) Seelampuri.

During 1975, Shri Jagmohan was Vice Chairman of DDA and Shri RS Gupta, a CPWD cadre officer, was Engineer Member. The planning of Resettlement Colonies was done by Shri RG Gupta Chief Town Planner of DDA. It was decided to implement this project on top priority. At that time, Shri MS Telang was Chief Engineer DDA and responsibility was given to him to implement the project. This development project was biggest project ever executed in Delhi. Shri RA Khemani, joined DDA as Superintending Engineer and was made responsible for Colonies to be developed in South and East Delhi. Both Shri MSTelang and Shri RA Khemani were CPWD cadre officers and on deputation to DDA. After some time, Shri Telang was transferred from DDA and Shri RA Khemani took over as Chief Engineer and responsibility for implementation of the total projects given to him. Dedicated team of Engineers of DDA worked for this project. It was decided that the work would be executed departmentally, by purchasing bricks from kilns directly. The earth for filling from specified locations and brick from brick kilns were carted by hiring private trucks. Shri Jagmohan, Vice Chairman, made administrative arrangements through Government of India that brick kiln owners and truck owners gave priority to requirement of DDA and would not refuse to work, if contacted by DDA Engineers. The work proceeded round the clock and engineers were ensuring that maximum resources could be utilized by employing adequate number of workers. In East Delhi, after a heavy pre-monsoon slower, there was water logging, because the area was low lying and proper outfall was not available. In fact, during a high level meeting in Govt. of India the project work was criticized with a remark that these colonies would need boats for transportation and roads were not required. Immediately thereafter, Vice Chairman arranged a meeting with Engineers as also visited the area. It was decided that level of these colonies would be raised by about 4 feet, to minimize the probability of flooding. For the source of earth, it was decided to dig the earth from adjoining area. This low lying area was developed as a lake and named "Sanjay Lake". A large number of trucks, coming to Delhi from other States were deployed for this huge task. Thousands of labourer were deployed for excavation, transportation and filling work. This huge task was completed in about four weeks period, before on set of monsoon. Besides, an outfall drain was constructed for draining excess water from Lake to river Yamuna. Area being low lying, pumps were installed and gates were provided in outfall drain. The work was completed before on-set of monsoons. The area did not get flooded during heavy rains. It was an impossible task accomplished in record time. The work done by Engineers was highly appreciated by Hon'ble Minister of Works and Housing.

Thereafter, the work of brick lining of drains and laying of brick Kharanja in lanes of colonies was started. This work was also taken up on a war footing. Most of the brick kiln owners in and around Delhi were persuaded for supplying bricks for these works on top priority to meet the huge requirement of bricks. This entire work also completed within a few weeks. These colonies were saved from flooding during monsoon. A commendable job of providing street lighting in these colonies was also carried out by Delhi Electric Supply Undertaking engineers. After completion of these works, colonies were visited by a number of VIPs during monsoons but there was no flooding. Sanjay Gandhi Lake stored most of the flood water and excess water was pumped out to Yamuna River through the main outfall drain.

Simultaneously with development of colonies, the removal of unauthorized encroachment from different areas was started and those who were living there were allotted plots in Resettlement Colonies. After allotment they were asked to shift to their new habitat. It was difficult exercise of shifting residence of lakhs of people to new locations, which hardly had any parallel in human history. DDA officers under the leadership of Shri Khemani ensured giving all the help and assistance on humanitarian grounds. It was all executed and implemented in perfect manner to avoid human suffering as far as possible. Within few years, these colonies were bubbling with happiness of people who shifted. They were satisfied with their resettlement. What a relief to poor citizens of the country, who became authorized owner of a house and started living in a dignified manner. There is no parallel of this project, in the world history.

II. SYSTEMATIC APPROACH FOR MAINTENANCE

In 1982, when KB Rajoria, a CPWD cadre officer took over as Chief Engineer (East) of DDA, all Resettlement Colonies of East Delhi were under his jurisdiction. These included Khichripur, Kalyanpuri, Himatpuri, Nand Nagari, Nand Nagari Extension, Gokulpuri etc. These colonies were already fully occupied and the responsibility to provide civic amenities and to ensure proper sanitation was that of DDA. This included water supply, scavenging of roads, cleaning of drains, maintaining public toilets and other services. The general standard of service was very poor. Strom Water Drains were full of mud and dirty water, giving foul smell. A number of toilets were chocked and there was blockage of septic tanks. Electrical bulbs for street lights were missing. At places drains and road needed repair. Hand pumps for water supply were not working at number of places. Residents were not happy with poor sanitation.

Review to work out plan of action- Residents wanted improvement in sanitation and DDA had workers, sweepers and supervisors for sanitation task. But residents were not cooperating and garbage was thrown in Storm Water Drains. Residents had pigs, cows and buffalos. All animals were spreading garbage and developing insanitary condition. Still residents wanted improvements and they suggested that defaulters who were creating insanitary condition should be punished. Workers employed by DDA were from the same section of the society as that of residents. Some workers were living in the colonies where sanitation was to be improved. They did not have any training for sanitation. Junior engineers and supervisors did not have basic training/experience for attending to sanitation work.

For job to be done by sanitation workers, it was ascertained from local bodies, as to whether there was yardstick. Besides, whether there were norms for frequency of services to be provided, materials to be used etc. Unfortunately, no authentic information was made available. Even for DDA Colonies, no standard norms of services to be rendered was established. In different colonies, different norms and standards were followed. After discussion with engineers and supervisors, norms for cycle of services to be rendered were developed, as brought out here.

(a) Civil Wing

(i) **Twice a day-**

(a) Cleaning of toilet blocks and public urinals,

(b) Water supply in hydrants from 6.00 AM to 10.00 AM and 5.00 PM to 8.00 PM.

(ii) **Once a day-**

(a) Scavenging of roads,

(b) Cleaning of subsidiary drains,

(c) Clearing of dust bins, refuse dumps and Dalaos,

(iii) **Once a week-**

(a) Cleaning of main roads,

(b) Cleaning of toilet blocks with phenyl.

(c) Cleaning of pipes connecting septic tanks of toilet blocks to open drains.

(iv) **Once a month-**

(a) Cleaning of WCs with acid,

(b) Providing new signboards, repair of damaged sign boards, replacement of missing sign boards and cleaning of sign boards including removing of posters etc,

(c) Checking of working of chlorinators and taking samples of treated water for testing,

(d) Cleaning of open areas and areas surrounding toilet blocks, dressing of road berms, filling of depressions including removal of broken electrical poles, dismantled materials unused building materials etc,

(e) Cleaning of Enquiry Office including stacking of materials in systematic and proper manner,

(f) Replacing broken WC pans,

(g) Repair of fencing of open areas,

(h) Replacement of broken RCC slabs over SW drains, (i) Replacement of missing parts and re-commissioning of hand pumps.

(v) **Once in three months-** Repair of damaged flooring in toilet blocks.

(vi) **Once in six months-**

(a) Painting of road marking on speed breakers,

(b) Repairing of compound wall and railing for the parks,

(c) Repainting of sign boards,

(d) Out-fall chockage clearance before rains and proper watch was to be kept during rainy season,

(e) Repairing of platforms for hand pumps and hydrants and waste water connection with drains,

(f) Repair and white washing of toilet blocks, parapet of culverts including repairs of damaged plasters etc,

(g) Cleaning of underground tanks,

(h) Repairs and repainting of water level indicator of overhead tanks.

(vii) **Once a year-**

(a) Repair and white washing of public buildings like pump houses, TV centers, Community Hall, Literacy Centers etc.

(b) Cleaning of overhead tanks,

(c) Cleaning of septic tanks.

(viii) **Once a three years-** Re-carpeting of roads normally once in 3 years or earlier.

(b) Electrical Wing

(i) **Once a day-** Giving TV Programmes

(ii) **Once a month-** Replacing broken and fused bulbs

(iii) **Once in three month-** (a) Checking and Servicing of electric pumps and ancillary installations.

Attending to incomplete, Repair & Corrective jobs-On the basis of complaints received from residents and during inspections it was observed that certain works were lying in the incomplete shape. For example, it was observed that on some roads only on one side the brick pitching was done. Such jobs were termed as incomplete jobs. Besides, there were certain structures, which needed repair. Such jobs were classified in the category of Repair Jobs. Moreover, it was observed that certain portion of the work was required to be redone for proper efficiency of sanitation service. For example, in some cases the Storm Water Drains did not have proper slope and needed repair and redoing. A list of incomplete/repair and corrective jobs is given below.

(i) Patch repairs of road cutting, pot holes, etc. on road with bitumen/brick pitching.

(ii) Repair of Water Supply distribution lines.

(iii) Repair of the SW Drains.

(iv) Correcting slopes in drains wherever necessary.

(v) Providing vertical grating at the outlet of subsidiary drains.

(vi) Rectification of defective invert levels in culverts.

(vii) Repair of Dust Bins.

The systematic approach for attending to cleanliness and maintenance, helped in improving the sanitation of Resettlement Colonies of East Delhi.

LESSONS LEARNT

(i) For such a huge project, it was absolutely necessary to pre-decide norms and conceptualize in the scope of master plan. It was done in excellent manner by Shri RG Gupta, Town Planner and his team. They deserves high appreciation.

(ii) For implementation of this project on priority, the patronage of Government, Town Planner of India and desire to implement, helped engineers to take decisions, beyond provisions of accounts codes and manuals. It was not possible to implement, a fast track project according to normal procedures. Shri Jagmohan, highly efficient administrator was Vice Chairman of DDA. It was his patronage and decision making skills, which ensured that project could be implemented. So impossible was made possible.

(iii) Consistent hard work by Engineers of DDA, under the leadership of Shri RA Khemani, ensured implementation of project on fast track. They deserve appreciation.

(iv) Norms for maintenance of Resettlement of colonies- No norms for maintenance of Resettlement colonies were available. It was initiative of KB Rajoria that such norms were developed and maintenance was done accordingly. It is an unique example of taking initiative to have systematic maintenance and give clean living environment to residents.

6

WATER RESEARCH CENTRE PROJECT AT BAGDAD (IRAQ)

Water and Power Consultancy (WAPCOS), a Govt. of India Undertaking, executed a number of consultancy projects in Water and Power Sectors in India and abroad. During late seventies, the work of conceptualizing, planning, designing, construction supervision and establishing Water Research Centre at Bagdad (Iraq) was assigned by the Government of Iraq to WAPCOS. This Center was to be similar to Central Water and Power Research Station (CWPRS) Pune. Accordingly, the design concept was developed by WAPCOS, with the technical support of CWPRS, Pune. After approval of design concept, WAPCOS engaged M/s Sachdeva Eggleston Pvt. Ltd. as Architects, for architectural and services planning of the project. They developed all the architectural and services drawings, for implementation of the project. The structural designing was assigned to M/s Mahender Raj and Associates.

According to provisions of contract with Government of Iraq, WAPCOS was also to provide supervision consultancy services for implementation of the project. As per the contract the project team was headed by a Scientist from CWPRS Pune. Besides, Senior Civil Engineers and Electrical Engineer were deployed at site to guide the project activities at Bagdad. Basically, this being a building project, WAPCOS requested Central PWD to provide services of Senior Civil and Electrical Engineers for this project. Accordingly, during 1978, KB Rajoria Civil Engineer and Shri SK Singhal Electrical Engineer were selected. Shri Lalwani, another Civil Engineer was already posted at Bagdad. All proceeded to Bagdad (Iraq) for ensuring implementation of the project. WAPCOS team was headed by Dr. SV Chitale, a Senior Scientist from CWPRS and after sometime Dr. GT Wadekar joined in his place.

At Bagdad, it was noticed that Iraqi people were friendly to Indians. Besides, the law and order situation of the country was very good. There was no difficulty in availability of food items. Besides, summers were extremely hot and dry whereas winters were very cold. This project was under the jurisdiction of Director General, Ministry of Irrigation and was coordinated by Shri Putrus, an Engineer in the Director Generals office. The project site was on the outskirts of Bagdad, on the bank of river Tigris. The area was low lying. The earth filling was done in consultation with project authorities. WAPCOS assisted the Ministry for preparation of detailed estimate and tender documents. Fortunately, CPWD Specifications, Delhi Schedule of Rates and Analysis of Rates were taken from India and these standard books were helpful in finalizing tender documents. Tenders were invited by Govt. of Iraq and M/s Engineering

Projects India Limited (EPIL), a leading Public Sector Enterprise under the Ministry of Heavy Industries, Government of India, was the lowest bidder. It was decided to negotiate with the lowest bidder. Engineers of WAPCOS along with Senior Engineers of the Irrigation Ministry, negotiated and finalized rates and special conditions. It was a good experience that Indian consultants along with Engineers of Government of Iraq negotiated with an Indian Construction Company. The project was awarded to M/s EPIL. Thereafter, Civil Works were sublet by M/s EPIL to M/s Hindustan Steelworks Construction Company Limited (HSCL). HSCL in-turn engaged private construction company. EPIL was responsible for coordination with the Iraqi client, consultants and all sub-contracting agencies.

The scope of work for the project included construction of buildings and other structures in the Water Research Centre Complexand maintenance for one year period. Structures to be constructed were: (i) Open hydraulics laboratory and pump house, (ii) Covered hydraulics laboratory, (iii) Soil Mechanics and Concrete Testing Laboratory, (iv) Administrative Building and (v) Auditorium. Internal and External Electrical works, Water supply and Sewerage System for the complex were also included in the scope of the work. For covered Hydraulics Laboratory as also for Soil and Concrete Testing Laboratory, Air washer system was provided for cooling during summer season. Auditorium had special features which included seating arrangements, simultaneous interpretation system and central air conditioning. All structures were of RCC frame and exposed brick work for walls. For the roof of Hydraulics Laboratory, Mero Space frame covered with sandwiched colour coated sheets was provided. The Mero Space frame covered large span without a column in between, as required for this laboratory. Viewing galleries were provided for observation and photography of models. Heavy duty concrete floor was provided. Open Hydraulics Laboratory and pump house had large area for model studies. Besides, huge quantity of water was arranged through pumping system for conducting model studies.

After award of work to M/s EPIL, the project team was posted at the project site by Ministry of Irrigation, under the Chief Project Engineer Shri Adai Hardan. Contractors M/s EPIL and their sub-contractor M/s HSCL also established their site offices. There was big team of Engineers with EPIL, headed by Shri VC Sharma, General Manager. Besides, a number of workers and supervisors joined the project. All of them were from India. Their residential arrangements were made at the project site itself and comfortable Caravans were provided for them.

The position of availability of building materials was reviewed with contractors. It was decided that source of locally available materials like sand and aggregates should be inspected and tested for suitability and quality. During inspection, it was ensured that testing was done, so that proper quality of materials supplied for the project. It was a commitment to quality from the source itself. Besides, most of the building materials were to be imported. For example, reinforcement bars, cement, structural steel etc were to be imported from Japan. The absolute control on quality of imported materials was exercised by Govt. of Iraq. Modern construction machinery was deployed at project site, which included Automatic Batching and Mixing Plant for concrete, Hydraulic Cranes with telescopic Boom, Transit Mixer, Concrete pumps, Cement Silos, Excavators and Hydraulic Rollers. These were also imported. Timely import of equipment and materials from other countries, which met the specifications, was necessary. Besides, for implementation of project it was necessary to have proper logistics towards arrangement of Indian work force, their continued presence at site, residential accommodation, and medical coverage.

WAPCOS, as consultants for supervision of the project, were the final authority for approval of material, levels, conformity to drawings for construction and so on. Only after acceptance of work by WAPCOS, Engineers of Govt. of Iraq were giving orders for implementation. At times, because of import of materials from different countries, specifications were different. Therefore, to examine and ensure suitability, code of practice of different countries were studied. Besides, to make implementation smooth, as required by Engineers of EPIL, additional structural drawings and details were supplied by consultants. On account of hard work by Engineers of contractors, the work proceeded according to program. Engineers of Govt. of Iraq were satisfied with the overall progress and quality of work.

During 1980, the Iraq-Iran war started. It became difficult to continue implementation of the project even though M/s EPIL made all possible efforts. In fact, living in Bagdad became a risky preposition. Fighter jets of Iran were moving at low height over Bagdad city. The project was practically stopped by M/s EPIL and efforts were made to evacuate and shift staff to India via Kuwait as there was no flight from Baghdad due to war. At that stage KB Rajoria returned to India and another Senior Engineer joined in his place. Of course, project team of WAPCOS continued at Bagdad till the project was stopped on account of war situation. Therefore, Government decided to stop the project till workable conditions re-established. It was a difficult task for EPIL to keep minimum staff at the site during war and therefore evacuation was done. All Engineers and other staff of EPIL returned to India.

After one and half year, the condition in Iraq became more or less normal. Therefore, M/s EPIL decided to restart the construction activity at the project site. At that stage M/s HSCL, the sub contractor, refused to remobilize, because it was feared that the Govt. of Iraq was not likely to pay for services to be rendered. Thus, it was for EPIL to reorganize and restart the work by taking responsibility to implement the project departmentally, that is, by employing work force directly. Besides, a bold decision was taken by M/s EPIL, must be in consultation with competent authorities, to implement the project by arranging funds from other sources. Simultaneously M/s WAPCOS also re-established its activities as consultant. But their stay was short lived because after six months the contract period of WAPCOS was over. Since planning and design work was already completed and construction activity was very much advanced, it was decided by Govt. of Iraq to close their contract.

Under financial constraints, M/s EPIL continued the implementation of the project and responsibility of implementation given to Shri Ramakant Gupta, Project Engineer. Project implementation in a foreign country was a big challenge, as this project was one of the early projects undertaken by any Indian construction company. In order to complete the work, EPIL engaged different specialized agencies, as also imported some machinery and equipment. Lifts were procured from Mitsubishi, Japan, the Air Conditioning equipment was imported from India and work was executed by M/s Voltas. The electrical work was done by M/s Crompton Greaves Pvt. Ltd, India.

EPIL completed the work in another one and half years that is by middle of 1984. Thereafter for one year, the maintenance work was successfully done by M/s EPIL as part of the contract. After a number of years Iraq Government repaid to Indian Government in the form of Sulfur and Oil. The Indian Government in turn paid in Indian Rupees to EPIL. EPIL lost substantially on account of stoppage of work due to war and remobilization, nonpayment of bills by Iraq Government resulting in heavy interest charges on borrowed money in foreign exchange and escalation of rates of all imported items.

LESSONS LEARNT

(i) **Readiness for new Opportunities:** Boom in crude oil prices in early seventies was a huge windfall for oil producing country like Iraq. As a result, large number of construction activities were going on with the participation of consultants, suppliers and contractors from several countries. Competition was therefore very tough. During this period, WAPCOS, a Govt. of India Undertaking, was operating at international level and taking projects with backup of technical expertise available in the country. The fact that WAPCOS could get this project and completed it shows that even during that period, Indian expertise was acceptable at international level. WAPCOS deserves appreciation for presenting the expertise in befitting manner and getting a project in a highly specialized field. Thus, developing and retaining core competencies always pays.

(ii) At that time EPIL, another Govt of India Undertaking, had developed expertise in project implementation and could get this project on merits under stiff competition. They completed the project under adverse circumstances, with the backup and support of Govt. of India. The initiative of EPIL is commendable.

(iii) **Project team selection and their engagement for the assigned work:** The organizational structure has to include agencies and personnel required for the job. This has to be properly determined and provided for. WAPCOS engaged external agencies/persons and made use of the expertise which was not available in house. It made use of the technical expertise of CWPRS, engaged architectural and structural planning firms, borrowed experienced officers from the CWPRS and CPWD to successfully execute the consultancy contract with Govt. of Iraq. Similarly, EPIL made use of its core competence and supplemented resources by engaging HSCL and other subcontractors to successfully execute the work. Heads of these organizations never felt shy to supplement what was not with them to have a robust team.

(iv) Workers, Engineers as well as Managers from India were appreciated in Iraq due to their skill, expertise and behavior. Indian Engineers engaged both by consultants and contractors did outstanding work. It speaks high of technical capability and expertise of Indian Engineers, who performed well even under adverse circumstance.

(v) **Creation and retention of Knowledge:** The senior engineers from India could prepare the estimation, design and specifications based on the book of specifications, and analysis of rates of Central PWD. Such publications are updated from time to time. It is considered a standard and relied upon by many construction departments and agencies. Thus, it is always helpful to invest manpower and time in developing and updating such documents as is being done by CPWD.

(vi) **Resource Management:** Resource planning is an essential element of timely project implementation. In this case availability and suitability of materials available locally, as well as from outside was examined at the beginning of the project. The specifications of materials to be imported was also studied to find their suitability and acceptance criteria determined in advance so that there was no mismatch.

(vii) **Risk Management:** In the face of adversity of war like situation, the EPIL did think it proper to complete the works even by arranging funds from their own resources, though suffered losses mainly due to interest burden resulting from delayed relief payment from Govt of India. It was more of fulfilling their commitment to honor the contract and not leave the client in time of distress. Such sense of commitment and responsibility was perhaps one of the reasons why the then Government of Iraq preferred to work with Indian companies, particularly government PSUs. However, it is prudent for construction companies be fully conversant with the issues of cost of financing through borrowings, to avoid severe losses on this account.

During late seventies National Building Construction Corporation Ltd(NBCC), a Govt. of India undertaking, was awarded a number of Civil Construction projects by Govt. of Iraq. A project office was established at Bagdad, Iraq. Shri P N Sadhu was the Project Director for Iraq works. The work for construction of a tourist hotel at Mosul in North Iraq was awarded as turnkey project to NBCC in August 1980 with period of completion as twenty months. The scope of work included all the Civil works, HVAC work, Electrical services, furniture and kitchen equipment, parks, road, parking spaces, finishing, and other miscellaneous work. This project was under the administrative control of State Organization for Tourism (SOFT), Govt. of Iraq. The lump sum contract value was about 8.75 million Iraqi Dinars, of which up to 80% was payable in foreign currency. The Indian Tourism and Development Corporation Ltd (ITDC), having experience in hotel industry in India, were the primary consultant to NBCC for the project. Back-to-back arrangement was also there with ITDC for providing and installation of items like kitchen and laundry equipment, carpet and wall hanging in the guest rooms, and a few items of furniture. The ITDC engaged sub-consultants for providing the design and BOQ for the building, electrical & HVAC, water supply and sanitary works, finishing items and horticultural works. The Architectural services were provided by M/S Jasbir Sawhney and Associates. Due to uncommon trapezoidal shape of the structure, the structural designs carried out by the M/S STUP India Ltd, consultant, were got checked by the client from one professor, Dr Hadid of Mosul University.

Soon after the award of work in August 1980, war between Iraq and Iran started and major conflict lasted up to June 1982. During this time flights to Iraq were totally stopped. This was a big setback to the mobilization for this project. When the intensity of conflict reduced, staff and workers could travel by air from India to Kuwait and then by road to Mosul. Adjacent to the site for the hotel, additional land was made available by the client for locating batching plant and locating/keeping other construction equipment, stores, temporary accommodation for the workers and office. Rented accommodation was provided for staff. One caravan, furnished with furniture and office equipment, was provided for the office of the site staff of the client in terms of the contract provision. During that period, Shri SC Kapur, joined as Chairman and Managing Director of NBCC. He provided necessary guidance and support in execution of the project. Shri OP Goel joined as Projects Director (PD) at Baghdad. Shri VG Ramdasi was the Regional Manager (RM), for Mosul as well as another hotel project at Dokan. Shri Biswas joined as the Senior Project Manager (SPM) at Mosul. The E & M works were initially coordinated by Shri SS Kapoor, and later by Shri Ram Ji Lal, Regional Manager (E) at Baghdad.

The hotel was designed as a twelve-story structure with Ziggurat type architecture (trapezoidal), as main elevational feature, in tune with the ancient Iraqi architecture of 2100 BC, built in the then Mesopotamia. Fire scape staircase on one end wall of the structure along with the architectural features on both the ends appearing like a special decoration, added to the beauty of the building. The covered area of 20,000 square meter translated into construction of 224 double rooms, 20 single rooms, 3 luxury suites, and one presidential suit. In addition, there were 12 Cabanas near the swimming pool, and 16 furnished staff quarters. Basement was to house bowling alleys, sauna, jacuzzi, disco with bar and fancy lights, squash court, and gymnasiums. A separate area in the basement was meant for housing services like AC plant, standby generator, water boiler, hydro pneumatic pumps, fire pump, stores and space for housekeeping staff. Guest entry was planned on the first floor by connecting elevated road from the city side to the reception foyer. The tall trapezoidal atrium from ground floor to the ceiling of the top floor in front of the reception specially added to the elegance of the reception area. The first floor, which was of double height, accommodated the main dining area overlooking the swimming pool, garden and the river Tigris. The first floor also accommodated shopping arcade, multipurpose hall and lounge, lift lobby, and entrance to the coffee shop overlooking landscaped fountains. The service area included the kitchen, control room for fire alarm system, staff dining, service lift etc. The mezzanine floor above the kitchen area served as the laundry room and other housekeeping activity. Remaining eight floors housed guest rooms along with some service areas for air handling units (AHUs), storage for housekeeping and the like. Rooms were provided on both the sloping sides. All the common areas were centrally air conditioned, air supplied through AHUs on each floor, whereas guest rooms were having individual fan coil units (FCUs). Music and sound system, MATV systems were also provided for public areas and individual rooms. The external area was designed to have a swimming pool with puddle pool, filtration plant all complete, Cabanas, two tennis courts, etc. Since the area was low lying compared to the adjoining road cum bund along the river Tigris, the feature was beautifully merged to convert it into an open basement. Area all around was for open parking and beautifully planned landscape with leisure walking pathways. An elaborate landscaped garden with pergola, walk-way to overlook the Tigris River was also part of the turnkey project. The roof of Cabana, a stretch of about 10 m wide and 50m long, *like baradari*, was for a panoramic view of Hotel Landscape and the Tigris River front. The ultimate finished hotel was to match a deluxe five star hotel.

Necessary arrangements were made, in the first stage as part of the mobilization, for arranging construction equipment, machinery, staff and workers. Considering the acceptance criterion of different works, the construction process was to be highly mechanized. Two tower cranes were erected, one on either side, which was necessary for concreting as well as carrying the construction material to different locations during construction. Practically all the equipment were brought on temporary import. Construction materials, meeting the specifications, were procured from countries giving best rates. Foundation for the structure was on cast in situ driven piles. Due to high Sulphate content in the soil, Sulphate resistant cement was used for all concrete works in contact with soil. As shape of the building was comprising two trapezoids trying to lean inwards, there were inclined structural members. Concreting of inclined members required utmost care. By careful planning and ingenuity, segregation while concreting was avoided. Slabs at ground and first floor were waffle slab and column arrangement to provide large open space without increasing overall height. Shuttering and placing of reinforcement in this area required highly skilled workers. Line, levels and finish achieved was excellent and were as good as final finish. The framed structure construction took about two years, on account of difficult condition of working during war period.

During April 1984, HK Srivastava took over as Resident Engineer(Incharge). At that time structural work was complete. Finishing works were about to start. Shri SP Baranwal, another Resident Engineer, was looking after Electrical and HVAC works. Shri Chakrabarti was Manager Finance posted along with supporting staff. The team was expected to complete the finishing work and commission the hotel for handing over in the minimum possible time. The first step was to ensure availability of required finishing material. The office of the Regional Manager, assisted by Shri SS Bhatia, prepared a list of materials required from the BOQ and drawings provided by the Consultants, and then obtained catalogues from potential suppliers. These were shown to client and after approval by the client, orders were placed. As these materials were not available locally, all the materials were to be imported. Sequencing of procurement was important from cash flow considerations along with progress of items of work. The entire work from client side was looked after by Ms Maha, Civil Engineer, one Surveyor and one Architect. For electrical works, electrical engineer was available. However, this small group kept full control on the works. Each and every material on arrival was checked with reference to specifications, method of fixing studied from catalogues, samples prepared and then only work was allowed to proceed. Many a times, the clients' Architect had his own idea about the joint thickness between tiles. Every day workmanship of the work done on previous day was checked and if it did not satisfy their requirement then there was no hesitation in ordering dismantling and redoing. So, it was a process of learning, making client learn and then workers to practice and do it correctly. Shri Hanna Salim, Client's Chief Engineer, used to visit the works from Baghdad office almost once a month. According to him, turnkey contract was to the turn the key, enter the hotel and find it functional in all respect. If the presentation drawing showed planter with flowers, planter and flowers are included in the scope of work. If swimming pool deck showed furniture and crockery, then these are part of the contract.

All the civil works were completed with workmen recruited by NBCC from India. Plumbing work was entrusted to M/S GSA Associates on labour contract. M/S Reunion, Mumbai were the electrical contractors. M/S Al Sharawi from UAE were the HVAC contractors (workers, engineers all were Indians). Lifts were procured from OTIS, France, its installation was awarded to M/S Al Nabooda. S/Shri Gopalan, MG Bhatia and AK Anand from ITDC looked after installation of Kitchen and laundry equipment. Installation of Sauna, Jacuzzi, Bowling Alley, filtration and chlorine dozing units for swimming pool were installed by the experts from the suppliers. Preparatory civil and electrical works were carried out according the detailed drawings made available in advance by these agencies. To increase the output per worker, 'incentive scheme' was introduced. Under this scheme, a group of workers were paid additional amount over and above their daily wage for the extra work carried out over a laid down yardstick. The cost had to be within the tender working cost adopted at the time of tendering. This scheme increased the output considerably and was preferable compared to payment of overtime which was generally without measurement of output. At the peak time a work force of nearly 300 workers were deployed almost round the clock. Problems associated with workers from different part of the country were required to be addressed timely. This was achieved efficiently by Shri Jeevan Singh, Office Assistant.

The floor area of 20,000 sqm was huge for the finishing work. There was scope to carry out works in each and every part of the project. The sequencing of the work was: a) providing service lines (water, sewage, electric, MATV, sound system, HVAC, fire alarm system, piping for the sprinkler system, supporting trays; b) completing cement concrete work;c) false ceiling, wall paper and delicate finishing work of fittings and fixtures; d) fixing of doors, door locks, placing of furniture, linen in respective areas were the last activity. On receipt of the material, each item required understanding how it was to be fixed by the departmental labour. Wherever required, Shri Ramdasi, the Regional Manager, took pains to explain the correct method of fixing/installation from the catalogues. If procedure was not clear, he obtained clarification from the suppliers. Telex was one fast mode of communication then. Thanks to the ingenuity and entrepreneurship of the Indian worker, workers became adept at adopting to new items of work in no time. Shri Kotru and Shri Thomas, AEs, guided the workers till the outcome was satisfactory. Nature of work was such that each area posed some challenge. Each room was to have wall paper finish over smooth plaster, cement concrete flooring with carpets, bathroom having concealed type flushing arrangement, bath tub, tiles on wall and floor. Same was the case with wall and floor tiles in public areas. The architectural finish showed certain thickness of the joints. To keep the joint thickness uniform special spacers were made by innovative method by our masons. The joints were to be filled by using specially imported mortar giving a grainy look (as in the catalogues).

Shri OP Goel, the Project Director, closely monitored each day's progress with the day's targets and provided required guidance for doing even more. During inspection visits he encouraged all, including workers, and provided immediate solutions. Shri Anil Kumar, the RE in Baghdad office, was very particular about timely submission of progress report each day. A few difficulties encountered are briefly mentioned:

(a) All the rooms were provided with carpet over cement concrete flooring laid in panels. When laid in panels, it was observed next day that corners curved up by a few mm, making the uneven flooring unacceptable. The issue could be resolved after a few trials by increasing the thickness of the flooring and keeping the surface damp after initial hardening in hot and dry environment;

(b) The client was very particular about waterproofing the toilet and kitchen area. Push fit type PVC pipes, imported from UK were used for the waste water disposal, whereas GI pipes from India were used for water supply. One inch thick glass wool insulation, covered with aluminium foil was used over hot water supply pipes. Pipes carrying chilled/hot water for the air conditioning also had insulation cover. The client insisted on checking each section of pipe line under pressure. After a section of pipe was laid, it was filled with water, pressure applied through hand pump, left for 24 hours. The joints were inspected for any sign of leakage, pressure gauge checked for any appreciable fall in pressure. If the dial indicated zero pressure, it was a sure indicator of leakage. In the process a few GI pipes cracked opened. Pipes had to be changed. The waste water pipes were filled with water, left overnight and checked for drop in water level, if any. After this step, floor of the bath was waterproofed, drains blocked, filled with water and floor below checked next day for any sign of seepage. All these were recorded by our engineers and certified by the client, like pour card. After tiling work was over bath room fixtures were fitted. End result was very satisfactory as no seepage or leakage was found during one year maintenance period after completion;

(c) Providing wall paper in all the guest rooms was another difficult task. From the catalogue, description of fixing wall paper looked so easy. When it was tried, Engineers struggled for a few days to get the method right so that there were no wrinkles, no gap between laps, no unevenness due to extra glue applied.

(d) Marble and tile were used as flooring and cladding in practically all the visible areas. Most of it was purchased from Italy. The marble (Travertine marble) wall cladding in lounge area was having pitted holes and trough on surface as if insects have eaten away and made holes in it. Stainless steel clamps were used for fixing. These needed delicate and expert handling as many pieces were breaking when cut to the required shape and size. Only very skilled workers, after training and testing, were used for such delicate works.

(e) During one of his visits, client's Chief Engineer enquired about the cladding proposing for the exterior surfaces. These were smooth shutter finish RCC surfaces and no treatment was envisaged by NBCC. The area involved was very large. The client brought out that as per contract all surfaces would be 'finished'. Besides, the architectural form is Ziggurat, was originally in mud bricks of pale-yellow colour. It was agreed to provide Sinjar stone cladding, available at a nearby place. These were fixed using Hilti bolt and stainless steel clamps especially imported for the purpose. With all our day and night effort it took more than six months to complete this activity.

Special features of Finishing Work

(a) All the public areas were to be provided false ceiling because according to contractual provision finished surface was to be provided. Besides, electric trays for electric cables, ducting for the air conditioning system were to run along the ceiling and these could not be left exposed. Mainly two types of false ceiling were specified in the architectural drawings.These were Gypsum based square tablets and metal strip Luxalon type. Both were to be suspended from the ceiling using Hilti bolts, suspenders and the supporting rails. Openings as shown in the drawings were to be provided for the light fittings, diffuser and air intake grill for the air conditioning purposes, speakers for the PA system and fixing of fire/smoke sensors. The entire finish, as seen from below had to be flawless and aesthetically pleasing. When the drawings were referred, it was noted that the prime consultant (ITDC) did not coordinate the drawings prepared by its sub consultants. First of all, the basement and first floor roof were waffle slab and not beam and slab or flat slab design. It had involved expensive plastic coated ply sheets to produce nice surface in the waffle which was to be ultimately covered up by false ceiling. The supply ducts for the AC services were having certain dimensions as per design, electric fittings having certain dimensions projecting inwards for bulb holder and the bulb were to be flush with the ceiling, and already procured. At several locations in the room, location of duct and electric fitting, were to come below the rib of the waffle slab. In turn the false ceiling would be less than 2m above the floor. This was not acceptable. A solution had to be quickly found. False ceiling materials, light fittings were already procured. As advance action, ducts were also made in sufficient length. It was necessary to provide the ceiling. Respective consultants were not locally available and when contacted were reluctant to provide a solution. It was time to evolve a solution at the project site. In different areas the location of the fittings and fixture as in the Architectural drawings were marked on the floor by chalk. Available fittings, duct size etc were back calculated and modifies pattern was made by adjusting location of fitting, diffusers, grills etc away from the rib of waffle slab. Then depth of the AC duct was readjusted, width increased to have the same cross-sectional area. Shri Handa and Shri Nair, AEs, contributed in this exercise and feasible patterns were created in respective areas. The client was also satisfied that no fixture or fitting is reduced from the original drawing. Besides, functionally performance was not affected and equally aesthetically pleasing.

(b) Doors to all the guest rooms were flush doors with two hour fire rating, and were procured from UK. Each room was to have locking arrangement. For hotels, locking system has to be designed based on security requirement as well as operational convenience of the hotel staff. The requirement was intimated in advance to the supplier for manufacturing locks and matching keys. Overall design of the locking system had to be kept in view while fixing locks to the doors. First requirement was the guest to have one key, which can open only the door to the room he is occupying. Usually, the key has a big and heavy key tag, so that the guest does not carry the key by oversight when leaving the hotel. It was the era before electronic lock and smart card keys. Next requirement was that for the house keeping staff – floor maid who is in charge of, say 20 rooms. Instead of 20 keys, the floor maid has one key which can open these 20 doors, but not any other door. Depending on rooms on a particular floor, numbers of maid keys are decided. Next is the supervisor, who may look after 2-3 floors. So, there is floor keys. Then supervisor key for a few floors. Then the master key which can open all the guest room doors. Last is the grandmaster key which can open all the guest room as well as all the doors in other areas. Thus, door locks had to be fixed matching with the sequence under which these were designed and ordered. These were carefully catalogued by Shri Narayanan (store keeper) when the supplies were received and issued, location wise for fixing. Care was taken that while fixing locks, the strip, fixed on the sides of the shutter, was not damaged. This would have resulted in damaging fire rating of the door.

(c) Activities outside the main building were no less hectic. A very elaborate landscape plan was prepared by the Architect and shown to the client. There were fountains, cascades, waterbodies, stone work, tile work, sloping ground, irrigation pipeline and lights. While the concept was excellent, there were huge financial and time implications, because the project team did not envisage it in their scope of work. For a contracting company, rushing against time, it mattered a lot. Once seen by the client, there was no escape. As a policy, CMD and the Project Director, both were of the view that NBCC was to do a good work and earn name not only for the NBCC, but also for India. After completion of stone/marble/tile works, horticulture works were undertaken. Added requirement was to use 'peat moss' mixed with earth in planters which had to be imported from UK. Also, in service area, a nursery including green house was made for propagation and upkeep of plants. It took almost one year for a team led by Shri S K Majhi, AE, to complete the work. Completed work was well appreciated by all concerned.

At this stage there was change in incumbency. Shri HD Sharma joined as the Project Director and Shri Ram Ji Lal joined as Reginal Manager (E & M).

(a) By middle of 1985, the completion of the hotel was in sight. Demobilization of workers also started to reduce overheads. It was time to think about commissioning the hotel by systemically testing the systems. This was taken up and all deficiencies were attended. Govt of Iraq decided to entrust operation of the hotel to M/S Oberoi (the famous Oberoi Chain) who were then already running one hotel at Baghdad. Their team, as operators, came to the hotel and went with tooth and comb in each and every area to ascertain what changes/improvements would be required to run the hotel smoothly as per their standards and requirement of local clientale. They were briefed about limitations as what NBCC provides would be within the Contract provisions. Still a long list was given and these items of work were executed – a few as extra items. Shri AK Bansal, JE, was specifically deputed to attend to these items of work. Their team made series of visits to see if anything more was missed and needs to be included. They were satisfied with our work.

(b) Once furniture items started arriving, client was requested to start the process of taking over the project. A taking over team was constituted by the State Organization for Tourism. As per normal practice in Iraq, the taking over team is entirely different than the staff posted during execution. The clients' site teams are equally answerable for the deficiencies as the contractor. The taking over process started in December 1985 and completed in March 1986. The team was very thorough and tested each and every item. That is to say each tap, each cistern, each door, lock, and even each smoke detector was tested for its satisfactory working. It also came up with deficiency list. The observations included even very minute details such as the shaft is not plastered and painted; location of the floor drain changed from that shown in drawing due to change in kitchen layout etc. Finally, all the work of the project was accepted by the team.

Two tower cranes were installed during construction for lifting and taking construction materials to different parts and heights of the project, and for concreting. Later on, these were used for taking finishing materials, fixtures (civil, electrical, HVAC) and furniture through the opening in the roof above the atrium. In addition, one material cum passenger hoist was provided in the atrium from the ground floor to the top floor during construction. As the finishing works and installation of fixtures neared completion, these were removed in phases, the last one to be removed was the material hoist. Carrying material or even walking to 8 floors was difficult. Closing of the roof opening was another big task. Considering it to be a minor matter, no detailing was done beforehand. It was considered sufficient to provide colourless Perspex sheet supported on Aluminium frame, with verticals at the ends. While it was planned to be fixed, the client came up with specific requirements that the material should be able to withstand ultraviolet rays and also there should be a safety valve type arrangement in case excessive pressure builds up inside the atrium (like in case of a blast). After some market study by Shri Ramdasi, such arrangement could be found and provided.

The Hotel project was finally declared as completed by Iraqi Authority and it was inaugurated in March1986. It was a proud moment for all Engineers of NBCC. Shri S C Kapur, the CMD NBCC came from India. After completion, it was named as Nineveh Oberoi Hotel. It was a five-star resorts overlooking Tigris River and number one destination for tourists. With pool, bar and Ferris wheel, this hotel became a favourite of Government officials and businessmen. Completion of this project was a major achievement for NBCC.

Due to prolonged war with Iran, and UN sanctions on Iraq, export of oil was limited. It was affecting foreign exchange position. Payment of foreign currency against work done was delayed initially, later became very scarce. In order to get funds released, NBCC was to coordinate with the teams of Export Import (EXIM) Bank, India and completed/ in progress projects were shown to them. Under this scheme, the Govt of India signed a contract with the Govt of Iraq to the effect that the money due to Indian contractors would be paid by Govt of Iraq in crude oil, as and when possible. In turn The Govt. of India would release payment to the Indian contractors in Indian Rupees with interest in due course of time. It also made available foreign exchange to pay for the purchases made abroad. This way payments could be made to suppliers and workers.

LESSONS LEARNT

(i) **Risk Management:** Working as contractor is a challenging and responsible task. While quoting for a work, scope of work has to be understood correctly and costing carried out accordingly. Lump sum tenders, based on vague details, unspecified finishes are likely to result in financial difficulties or constrains which may affect progress of work. It can also lead to loss of reputation if not financially supported by other sources. The scope of work involving different consultants requires proper coordination and cannot be left to a party who is not financially affected by the decisions. Fortunately, all the coordination work was done in effective manner.

(ii) **Management Responsibility:** It was decided by the top management that quality work has to be done in least possible time. The reputation of the Country as well as the Corporation were involved. This gave sufficient encouragement and incentive to those working for this objective. It made good business sense too. Good reputation helps in getting more work to a contractor, and client tends to consider extension of time, variation orders keeping in mind positive actions.

(iii) **Improvement in process / products and related learnings:** One need not be shy of using new techniques or materials. There is no limit to human ingenuity. Example is the 5 star super deluxe hotel created by Engineers using many new materials and technique with success.

(iv) **On-the job Training and Skill enhancement:** Training of workers to use new equipment, technique and materials is essential. In the present case, Engineers took initiate to train workers which helped in completing the project successfully. Determination to succeed and self-confidence, backed by coordinated hard work are the *mool mantras* for success of this project.

Hotel at Mosul (Iraq) -
Front Elevation of the Hotel

Hotel at Mosul (Iraq) -
Another view of the hotel – staff quarters, multipurpose hall
and the guest rooms

Hotel at Mosul (Iraq) –
Coffee shop upper terrace

Hotel at Mosul (Iraq) –
Outside elaborate land scape with fountains, cascades, lights

8

WATER TREATMENT PLANT AT DIBIS (IRAQ)

During 1982, the work for construction of a Water Treatment Plant (WTP) at Dibis, near Kirkuk, was awarded to National Building Construction Corporation Ltd (NBCC) by the Oil Ministry, Govt. of Iraq. The awarded contract price was 1.75 million Iraqi Dinar out of which up to 80 % was payable in foreign currency. At that time armed conflict with Iran was still continuing.

Shri SC Kapur was the Chairman cum Managing Director (CMD) NBCC at Delhi. Shri OP Goel was the Project Director (PD) for Iraq projects and Shri SS Kapoor was the concerned Regional Manager (RM). Their office was located at Baghdad. HK Srivastava joined NBCC on deputation from CPWD and was posted as Resident Engineer (RE) for the WTP project at Dibis. The Projects Director gave two important advices for working as a contractor. These were i) it was always to be remembered that the junior most person from client was senior to the senior most person in NBCC; and ii) Project in-charge should ensure receiving regular payment from client to ensure proper functioning of the organization. The Resident Engineer (RE) was assisted by two Assistant Engineers Civil, one Assistant Engineer (Electrical), one Junior Engineer, one Electrical Foreman, three Accounts Assistants, one Doctor and one Office Assistant. Shri Basu Malik, a seasoned AE from another nearby project at Mosul was deputed to this project for a short term to assist in establishing the Project Office and familiarize the newly posted Engineers with procedural requirements.

Dibis was a small village, about 40 km away from Kirkuk, an important town of North Iraq. The project site was along the bend of river Little Zab, a tributary of river Tigris. Near the project site, the client provided land for establishing temporary residential accommodation for the workers, staff, stores, and place for keeping and carrying out maintenance of equipment. As the camp was not yet set up, it was not possible for anyone to stay at site on account of extreme heat. At the same time, people were required at site to create basic facilities. To start with, a base camp was set up at Kirkuk where NBCC was executing another project. Everyday staff moved to the site with workers and carried out required work. Items like GI pipes, electric switches, cables were purchased from Kirkuk. Caravans were imported from Kuwait to provide accommodation at project site. Thereafter, air coolers and sanitary fittings were provided. For Kitchen and common room precast concrete hollow block walls covered with CGI sheet roofing with insulated false ceiling were provided. This gave larger area and ensured protection against fire. Once a few caravans were ready, camp was shifted to Dibis. The work area and living accommodation was contiguous. Further mobilization of men, material and machinery continued as required.

Mode of working was to carry out work departmentally by deployment of men and machinery and purchase of materials. Workers were recruited by NBCC in India and brought to project site. Most of the workers could adjust, over a period of time, to difficult working conditions. Vehicle drivers from India had to be especially careful as it was right lane drive in Iraq and the average speed was around 100 kmph. Initially, there were a few accidents, luckily non-lethal. In cases of accident Iraq followed the 'Hammurabi code', i.e., 'an eye for an eye'. After the accident the driver was kept in jail till the injured person was cured or his relative told the police that that they had nothing against the driver. Cost of medicines was also paid. In one specific case, the Resident Engineer went with interpreter to the house of the injured person to request and get the driver released from jail with their consent.

Practically no construction material, equipment, spares etc were available in Iraq. Everything was to be imported from other countries. Plant and machinery were to be brought on temporary import. On completion of the project these were to be exported back or transferred to some other active project. Spares required for Plant & Machinery were also not available and imported. Imported equipment were of state-of-art products. With some guidance and patience, workers got used to working on new equipment. This added to their efficiency. Procurement of equipment, material had a long lead time and was planned in advance so that there was no idling of workers. At the same time cash flow issue was to be kept in sight.

The designer was a French consultant who provided all the structural details and BOQ. Complete details were available before start of work. Detailed specifications were also available. On study of project drawings, it was noted that though the name of the work in the Contract was 'Water Treatment Plant', the project was that for a river intake structure, pump room with gantry girder, foundation for pumps, foundations for supporting water pipes on the sloping hillock leading to the powerhouse area and the actual treatment plant, a double height shed with steel structure and CGI sheet roofing to house secondary pumps and controls. Water was to collect from the river in a storage chamber by gravity flow. The storage chamber itself was at two levels. Nearly half portion of the storage tank towards the river was about two meter lower than the one toward pump side. It was to act as collection chamber for silt and debris. Iron gate with grating was provided to prevent entry of floating materials, large sized debris etc. Roof of this chamber was little above ground level. The pump house was for housing six heavy duty pumps to pump water from the collection chamber to the treatment plant. It was of double height. RCC columns were also provided to support gantry crane for installation and replacement of pumps. Sleeves were provided in the pump room walls at the time of concreting for connecting suction pipe in the collection chamber and connection to the pumps on the other side.

All the concrete was to be produced only with weigh batching. The manner of transportation and placing was given in specification. Based on drawings and schedule of items to be executed, requirement of equipment, shuttering, materials etc was sent to the Regional Manager (RM) at Baghdad. Whatever could be spared from other project sites, was shifted to Dibis. For remaining quantity of machinery, equipment, materials, orders were placed for import. In the meantime, immediately required equipment was hired from the market till the equipment ordered for this project could arrive. Shri SC Kapur, CMD, NBCC also visited the project site and assured for immediate procurement of equipment and materials as required. Iraq Govt. working hours were from 7 am to 3 pm. But, by mid-morning during summer temperature used to reach above 40^0 C and it was difficult to work in the open. The Project office of Oil Ministry was 40 km away from site at Kirkuk. Shri Hasan and Shri Zaki were Project Engineers. They were able to communicate in English. It was directed by them that at every stage, they would check the work before allowing concreting and other construction activities. During their visit to site, with prior arrangement, the checking was done and jobs (pour card) got approved. Thereafter in summer season, most of field activities were carried out early in the morning or late evening/night. Winter was quite harsh and no night time activity was feasible. This way work on foundations for the pipeline and for the steel structure was executed under efforts of Shri John, Assistant Engineer.

By using Excavator, Loader and Dumper, the excavation for the intake structure was started. Immediately thereafter subsoil water from the adjoining river was met. It was pumped out using suction pumps of 5 HP capacity. As excavation went deeper, pumping was found to be inadequate. After the situation was studied, the Regional Manager contacted a firm specialized in 'well point' system. After negotiations, a contract was awarded for installation of well point system for dewatering. When they started the work, it was expected that this system would work. However, as the soil was clay mixed with pebbles, the 'well point' system could not work. In the meantime, excavation using suction pumps was done, but without much success. One D-8 Bull dozer was brought for diverting the flowing river away from the project area, by making a bund between the river and project. It reduced flow, but still it was not as per requirement. It looked that we would not be able to implement the project in this manner. During his visits, the Project Director encouraged the team and promised to give all possible support and guidance. Meanwhile Shri AK Kaul joined in place of Shri SS Kapoor as Regional Manager. Solution came unexpectedly during visit of Shri Narain Das, Assistant Engineer, to nearby UP State Bridge Corporation site where work for on construction of a bridge was in progress. On knowing about the problem at Dibis site, Shri Mittal Project Engineer UPSBC, offered to loan submersible pumps of high capacity, which were used for dewatering for construction of well foundations. Two such pumps were brought on loan and installed on the same day. This arrangement worked. Thus, it was possible to lower the water level and carry out the excavation. Excavation was done day and night without difficulty. This was the breakthrough for this project. The Regional Manager was informed of the success. After his visit to the site, it was decided to get more such pumps of higher capacity and order was placed. The excavation up to the desired level was completed by using these submersible pumps and by diversion of the river. Determination of the staff, of course, was the positive factor. Shri Rana, Foreman Electrical was very particular about keeping pumps in operation all the time. Clients were very happy and were hopeful that the work could be completed at an early date.

Next step was to plan for the concreting of two meter thick raft foundation which was to be completed in one continuous operation. It was estimated that with the batching plant installed at site having capacity of 25 cum per hour, it would take little more than 30 hours. The Project Director, during his visit gave necessary guidance and also insisted that there must be standby additional capacity available for concrete batching plant, transit mixer, concrete pump, vibrators and generator. A duty roaster was prepared. Manpower and machinery were borrowed from adjoining project at Kirkuk. Two concrete pumps- one stationary and one mobile were deployed. Once all arrangements were in place, concreting work was started early in the morning. A number of Engineers from the client side had come. After about four hours of concreting, one of the moving machineries ran over the wire supplying electricity to the submersible pumps and these pumps stopped working. Soon some water seeped into the concreting area. The Client immediately ordered the concreting to be stopped. After consulting senior engineers from Client, it was decided that already laid concrete should be taken out as it could have become honeycombed. Thereafter it was a tough job chipping away the rich concrete with embedded reinforcements. It was a hard job to be done. It took nearly ten days to take out the concrete using jack hammers and chisels, relay the reinforcement and shuttering. It was also ensured that this time there was no tripping of electric supply to the pumps or any other equipment. Accommodation was provided to the supervisors from the client to stay overnight. Thereafter the concreting done under dry conditions with full compaction. The concreting of the raft was a success and the main hurdle, physical as well as psychological was crossed. The pumping had to be continued till the roof was laid, so as to add sufficient dead weight and the structure does not float like a concrete boat. Shri SC Kapur, CMD, NBCC again visited the project. His soft spoken but firm words were full of encouragement and enhanced confidence of the team.

Next challenging task was construction of the Pump Room. The Pump Room was of double height, with arrangement for gantry crane. Acrow (trade name) props were used for staging. The ceiling height was such that two props, extended to their full height were required to be kept one over the other. It was something which was not executed earlier. Acrow (trade name) UK, when contacted by Shri AK Kaul, Regional Manager, did not have a solution to offer. Everyone considered it a risky proposition. Especially due to vibrations on account of occasional anti-aircraft guns firing from adjoining hillocks. Available literature was studied. Ultimately, it was decided to follow common engineering practice and go ahead. First, lower layer of staging including bracing was erected. Horizontal pull was applied using wire rope and excavation machinery. Once found stable, then wooden sleepers were placed and firmly fixed. Over this layer of sleepers, second stage shuttering was erected with cross bracing and the shuttering for beams and slab completed, reinforcement placed and concreting carried out using concrete pump. Almost 72 hours passed without any distress. Everyone took a sigh of relief. After the shuttering was removed and staging taken out Iraqi Engineers appreciated the challenging task accomplished. Of course, Shri Narain Das, an experienced AE from CPWD, stood firmly during the entire operation. Thereafter remaining works were completed. The contract work was declared completed and taken over by the Client on 30.04.1984. Full extension of time was granted without any penalty and all payments made.

Later on, it was informed by head office that the expenditure for the project, including Baghdad office expenses was to the tune of 53 % of the payment received for this contract. It was a huge profit. This was primarily due to the fact that while the mobilization was on, the works which could be undertaken were started. Besides, duration for the depreciation charges for the plant and machinery and shuttering was for a lesser period compared to time for implementation, as work was done very fast and at times round the clock. A few months lost in finding the correct method for dewatering, were more than made up by extra hours of working. All the workers and the staff deserved compliments for the same. They were of course suitably rewarded by the Corporation for their hard work.

LESSONS LEARNT

(i) **Understanding the Client's requirements:** Working in a foreign country and a different organization required adjustment and understanding the ground rules. The performance was outstanding. Thus, in turn the outcome was extraordinary.

(ii) **Use of modern equipment:** Adoptability to new Skill is one special trait of Indian workers. With some hand holding and guidance they picked up use of new equipment and construction methods. In turn it was successful completion of the project. Training is an important feature of successful projects.

(iii) **Resource Planning:** Resource planning and mobilization are key elements, particularly each procurement has a long lead time. At the same time, a contractor has to keep in mind that cash resource is not blocked.

(iv) **Learning from Failure:** Client stopped concreting as water seeped into the concreted area and asked for chipping. This unforeseen situation was not imagined and in turn dismantling of incomplete concrete work was done. Proper and appropriate tools and machinery were used for achieving the objective. Ancient wisdom from poet Rahim also says that there is no use of a sword where needle is required. This was put into the practice by the Project Team. There was saving in cost incurred and in turn profits were very high. It was a highly successful project.

Water Treatment Plant at Dibis (Iraq) –
Base concrete completed for lower and upper portions of the storage chamber

Water Treatment Plant at Dibis (Iraq) –
Work on Pump room walls in progress, sleeves for the suction pipes are also seen

Water Treatment Plant at Dibis (Iraq) –
Construction activity on pumproom - during night

Water Treatment Plant at Dibis (Iraq) –
Double staging for the pump room

9

REGULARIZED UNAUTHORIZED COLONIES OF EAST DELHI

After independence of the Country in 1947, a large influx of people migrated to Delhi from newly formed country Pakistan. Thereafter, the emergence of unauthorized colonies started. Up to 1974, as per the policy of Government, 471 colonies were regularized. Thereafter from 1974 onwards, on account of gradual inflow of people from adjoining states and other places, unauthorized colonies, gradually developed in geographical limits of Delhi. These colonies developed in vacant land, low lying areas, agriculture fields as also some Government acquired land. Permanent or temporary buildings were constructed in these colonies after leaving space for roads, drains etc. These colonies were developed without authorization of Delhi Development Authority and in complete disregard to masterplan regulations. Basic amenities like roads, water supply and electricity were not provided to residents of these colonies.These settlements were not legal in real term. But from human consideration as also from social and political point of view, it was considered necessary to give some relief to residents of these colonies. During rainy season, on account of poor drainage, the living was difficult. Residents of these colonies approached Government and to make living comfortable, earth filling and dry brick pitching was done in some of these colonies. The area wise distribution of 150 unauthorized colonies, identified by Government at that stage was (i) East Delhi (the area across river Yamuna)-87, (ii) South Delhi-32, (iii) Outer Delhi-13, (iv) Karol Bagh-12 and (v) New Delhi-06. The total area of these colonies was 4500 hectare and about 12 lac people were involved. These colonies were regularized by Government vide Resolution No. 116 dated 29-10-1984. At that time Shri HKL Bhagat was Hon'ble Minister of Urban Development and he was member of Parliament from East Delhi.

During that period, Shri KB Rajoria was Chief Engineer Delhi Development Authority for East Delhi area. The responsibility for providing minimum basic services was given to him. After preliminary survey and investigation, it was noted that most of the colonies were not having proper and adequate drainage system. Residents of these colonies approached the Government for giving immediate relief by earth filling and brick pitching on roads. They were informed that since colonies were regularized, the DDA would give permanent relief and not temporary relief. They were requested to cooperate with DDA Engineers. They believed and cooperated all along.

The drainage system of Trans Jamuna area was studied. It was noted that this area was lower than HFL of river Yamuna. On river side embankment of adequate height was existing which was higher than HFL of river Yamuna. So, no flooding on this area was possible, directly from river water. For drainage of area drain number one and two were existing and at outfall location proper Gates were provided. One more drain, parallel to embankment, was also existing. Normally during dry season, the gate was kept open and the outflow of waste water from drains was possible by gravity. During floods, the gate was kept closed and waste water was pumped to river Yamuna. This system of drainage was working properly and nicely. For this project, the drainage system of unauthorized colonies was to be connected to existing drains.

On the basis of geographical location regularized unauthorized colonies were divided under four groups. These were (i) Laxmi Nagar Group, (ii) Shakarpur Group, (iii) South side group near National Highway and beyond and (iv) North side group including Mauzpur and Ghonda. The purpose was to plan development activities and drainage system, for these groups, separately.

Laxmi Nagar Group of Colonies- These colonies were on the north side of Vikas Marg and towards east of Yamuna Bund. For outfall drainage, a disused canal was found on north of this area. This canal was part of drainages system and connected to Drain number one. Unfortunately, the level of disused canal was higher than Laxmi Nagar. It was decided to lower this canal, so that water could flow by gravity from drains of Laxmi Nagar. Thereafter levels of main drain of the colony and drains along roads of the colony were finalized. The level of road was decided to ensure that water could flow from road to drains. It was further decided that work of drain and road should be done simultaneously and included in one contract. Thus, contract packages were divided into three parts, one for outfall drain and disused canal and other two for road and drains of the colony. The work for lowering of disused canal and outfall drain was started, before undertaking road work in colony but the residents were not happy. They wanted that the road opposite their houses should be developed, before drain work. Consequences regarding flooding in the colony, without outfall drain were explained and they were satisfied by DDA's programming. Thereafter, the work was started. It was very difficult to construct road and drains in inhabited area. Fortunately, public was fully convinced with DDA's planning and programming and everybody cooperated. After completion of work, during rainy season, it was noted that drainage was proper. Residents of the colony were satisfied and happy.

Shakarpur and South side Group of Colonies- Vikas Marg is towards north of these groups of colonies and National Highway-24 Bye-pass towards south. Towards west is Yamuna Bund and an outfall drain flows adjacent to the road along the Bund. Fortunately, these colonies were found to be higher than the drain. The drainage system of these colonies was designed properly along roads and connected to outfall drain. Thereafter the work of roads and drains was awarded and properly executed. After completion the work these colonies were having proper roads and drainage system. There was no flooding of area.

Mauzpur- Ghonda Group of Colonies- These colonies are near Yamuna Bund. Towards north is Wazirabad Road and towards south is old GT Road, connecting Delhi with Steel Railway Bridge. Towards East is main road and main outfall drain is along this road. After detailed survey of these colonies, it was concluded that, drainage system was possible in different colonies. But on account of dense habitation there was no space for constructing the interconnecting outfall drain. Efforts were made to design and develop the main outfall drain along the approach road. But this road was not having space for accommodating main drain of adequate capacity. Therefore, even after developing roads and drainage system of individual colonies, capacity of main drain was not adequate to provide proper drainage to the area. One possible solution considered at that time was, to develop main drain after dismantling the main road and thereafter fully cover the drain and use it as road. For this extra work, approval was not available. Therefore, drainage problem of this area could not be solved. Thus, inspite of the fact that roads and drains were developed in different colonies, the drainage problem remained unsolved.

It is a pleasure to recall, experience of working under difficult situation to solve problem of residents and give relief. Depending on situation and after detailed study, issues were resolved by innovative solutions. In a broader sense, it was service for the society.

LESSONS LEARNT

(i) **Empathy by Government:** The concept of Regularized Unauthorized Colonies was innovative and suited to reality of life. Those who came to Delhi in search of a job and after getting job, started living in Unauthorized Colonies as they had no other option available. The concept of regularization of these areas was a positive bold initiative by the Government. It had no precedence. Thus, these who took this decision, did a commendable job and their efforts are worth appreciation. Those who settled in these unauthorized colonies, really got benefit beyond their imagination and expectations once these colonies were regularized.

(ii) **Evaluation of Risk:** There were negative consequences for the future. Unscrupulous elements encroached upon or purchased agriculture land and developed unauthorized colonies for their commercial benefits. They approached Government for regularization and the trend continued. Thus, they took undue advantage. It is not a proper trend and should be discouraged.

(iii) **Planning for restoration of Quality of Life:** Adequate planning and designing are important in taking up any remedial measure. In the present case requirement was to take the storm water to the disposal point, i.e., the river. The desire of the residents was to have proper roads so that they may live comfortably during the rainy season. Providing the paved road gave relief as the stagnant water would have damaged the road. Thus, providing the proper engineering solution worked very well.

(iv) **Involving end user:** Residents of these colonies were always informed about works to be executed and reasons for prioritization. Works were executed only after residents were satisfied with details of programming. They were happy after works were executed. A number of years after these development works were implemented, Shri KB Rajoria visited some of these colonies and enquired about drainage system. Residents informed about satisfactory working of drainage.

10

QUALITY SYSTEMS FOR VASANT KUNJ HOUSING PROJECT AT DELHI

During eighties, Under Vasant Kunj Self Financing Housing Project of DDA, about 2000 Self Financing dwelling unit of two and three bed room category were to be constructed. This project was under the jurisdiction of Shri RA Khemani, Chief Engineer South Zone. For this project, a number of packages were made and work was awarded to different contractors. After award of work to contractors, when the project was just started, KB Rajoria was transferred as Chief Engineer, in his place. This was a prestigious project. Therefore, in consultation with Engineers and contractors, it was decided that to ensure total quality, a systematic approach should be evolved. It was also decided that for this purpose procedure for checks and manner of documentation should be framed. The purpose was that checks during process should be well documented and impact of quality should be felt by allottees. Accordingly, systems were evolved, implemented and documented.

Systematic approach for Quality – Various steps taken to establish systematic approach for testing of materials, to achieve quality standards are as follows:

(a) *Establishing Testing Laboratory at the project site* – It was considered desirable to establish a field laboratory for the whole project. This laboratory was having facility for testing of materials like concrete, bricks, aggregates etc. Following testing equipment were provided in the laboratory: (i) Hydraulic Compression Testing Machine, (ii) Aggregate Impact Value Test Apparatus, (iii) Sieve Shaker and set of Sieves for fine and coarse aggregates, (iv) Physical Balance of 5 kg capacity, (v) Spring Balance of 100 kg capacity, (vi) Platform balance of 300 kg capacity, (vii) Vicat needle apparatus with dashpot, (viii) Meter to measure moisture in timber (Range 7% to 40% moisture), (ix) Electrical oven with temperature control - range up to 250°C, (x) Water distillation Apparatus, (xi) Cube Mould 15x15x15 cm, (xii) Vernier Calipers, (xiii) Screw Gauge, (xiv) Smoking Machine, (xv) Pressure Pump. The consolidated list of tests to be done was taken from CPWD Specifications and its copy made available to site staff and contractors.

(b) *Testing of Materials* – Testing of materials was done at the project site laboratory. It was ensured that materials were tested before use. Only after test results found satisfactory, materials were allowed to be used. A few representative samples of the materials were also sent to approved testing laboratories. Test results from approved Laboratories were compared with test results of site laboratory, to ascertain correctness.

Control of different contract packages – Under different contract packages, work for a few blocks was awarded. Each block had two dwelling units in each floor, making eight units in a block. For example, a contract package of 120 dwelling units had 15 blocks of eight dwelling units each. Stages of construction of each block were divided as follows: (i) foundation work upto plinth level, (ii) brick work and RCC work up to Ground floor lintel level, (iii) brick work and RCC work up to 1st floor level and so on up to RCC work at Terrace level. Beyond this level, all work was considered at terrace level. Stages of different blocks were correlated with materials brought at site. Test samples were taken at different stages of work of different units. Thus, test sample was identified with the particular locations in each block.

Record of Laboratory test results of materials – The record of all the tests for one particular contract was kept in a Master Register, instead of several registers. The block number and stage of work in the block was recorded against each test. Master Register had other records like date of issue of drawings, Cement register, Steel Register, Site Order Book, Hindrance Register etc. The frequency of tests was pre-decided according to requirement of CPWD Specifications or relevant IS codes. Thus, it was ascertained as to how many tests were required for a unit at different stages. The programme for different tests was accordingly decided. From this data, it was possible to know the date on which test was conducted for material of a particular location in the specific block. Proformas were developed to record the data as per block number, and stage of work. For each unit at different stages, it was possible to work out number of tests to be undertaken depending on quantities of material consumed. In order to review the data about number of tests required and conducted, a Material Test Review Proforma was developed. In this Proforma for each material, the quantity of materials brought to site during a month along with number of tests, required to be conducted and actually conducted were recorded. It was also mentioned as to whether test results were as per requirements of specifications or not. Thus, if material was not fulfilling requirement of specification, it was possible to reject it, before using it for the construction work.

ISI Marked Materials – Only ISI marked materials, if available, were used. For such materials, only conformity tests were done. As an extra precaution, the source of supply was also recorded. Normally, specific brand names of materials mentioned in the Contract Agreement were used. Still required tests were conducted.

Workmanship – Proper workmanship was possible only by employing capable, well experienced and well trained workmen. Strict watch on workers was kept and workers not found suitable to produce quality work were removed from project. It was the responsibility of field Engineers to ensure that proper workmen, i.e., masons, carpenters, bar benders, plumbers, electricians etc, employed by contractors. Besides, contractors were required to keep record of work done by individual skilled worker or a team of workers. During inspection, whenever sub-standard work was noticed, workers who did the work were identified. Suitable action was taken to remove such workers. Besides, if ITI trained workers and supervisors were available, they were employed. Steps were taken to verify competence of supervisory staff of contractors. Only these persons were allowed to work, who were competent and had adequate experience.

Structural work Inspection – As already mentioned, the structural work (brick work and RCC work) was inspected after completion of work at different stages. The defects/deficiencies up to or for that stage were recorded in Review Proforma. Thereafter, these defects / deficiencies were rectified and recorded in Review Proforma. After rectification of defects/deficiencies each block of eight houses was inspected by Superintending Engineer. Observations by SE were intimated to the Contractor and defects/deficiencies if found were attended. Thereafter, work was allowed to proceed further. This approach of systematic inspection ensured that structural work when completed was without any defect.

Services Inspection – The work of internal services was allowed only after structural work was completed and defects/deficiencies removed. Work undertaken for internal services included water supply lines, drainage pipes, sewage pipes, electrical conduits and telephone conduits. These pipes were laid by cutting chases and by making holes in the brick work. After completing these works all services were checked to ensure that functional requirements were fulfilled. A very important decision taken was with regards to drainage and sewage pipe lines. Instead of right angle bends, bends with one hundred twelve and half degree angle were used. These bends were specially got manufactured. By using these bends, possibility of leakage in drainage and sewage pipes was reduced to almost negligible. Water supply lines and electrical conduits were checked by pumping water to ascertain leakage if any. Sewer lines and drainage lines were checked with Smoke Machine. By and large all test were according to requirement of CPWD specifications and corresponding Indian Standards. Defects got rectified and services work was certified as completed in each block, before proceeding with finishing work.

Final Finishing – After services were found in order, other works like plaster work, flooring and finishing jobs were taken up for each unit. After completion of work for each block, it was inspected by Executive Engineers (Civil) and (Electrical) as also representatives of Housing Department and Architects. All defects and deficiencies pointed out by them, such as reverse slope in toilets and kitchen, unfinished corners, defective plaster work, defective flooring work etc were attended. Besides, doors and windows were also fixed simultaneously and also checked for defects if any. These defects were attended immediately thereafter.

Stage Checking Records – Stage checking records were kept in systematic manner. These were signed by Contractor's representative and Junior Engineer/ Assistant Engineer. Thus, a report was submitted after completion of work of each stage, in every block from foundation work to completion. These reports were having following information certified by Contractor and Assistant Engineer: i) All dimensions are according to structural and architectural drawings and there are no variations. Variations if any were recorded with reasons, ii) The quality and geometry of work was checked and found according to requirements, iii) Centering and Shuttering provided for RCC work was checked and found safe and proper, iv) Reinforcement for RCC work checked and found to be according to structural drawings, v) Laboratory test results for work under consideration were found to be satisfactory, vi) Electrical conduit work and pipe lines for water supply, drainage and sewerage checked and tested and found proper and without defects or leakages. It was ensured that next stage work, started only after the earlier stage work was checked and defects rectified.

External Services- For the purpose of control on quality of external services, pockets of about 50 to 60 blocks were considered as one unit. In order to implement the work of external services in systematic manner, a Combined Services Plan was prepared. This Plan was having details of roads, water supply lines, sewer lines, drainage lines, electrical cables, telephone cables etc. Besides, location of existing trees was also marked on this Plan. The reduced level of all the services was also shown in the Plan. Cross sections at different location were also given. In addition, detailed drawings were made for all the locations where services were crossing one another.

Testing of Services- All the services, i.e., water supply, sewerage, drainage, electrical and telephone lines etc were tested as required. These services were covered only after test results were found satisfactory.

Final Inspection by Senior Officers - After the work for a pocket including internal services and external services was completed, a joint inspection was conducted by Superintending Engineer (Civil), Superintending Engineer (Electrical), Executive Engineer (Civil), Executive Engineer (Electrical), Dy Director of Horticulture and representatives of Housing Department. Defects noticed were got rectified and thereafter, project declared completed. In this manner work for all the pockets was completed. The finished project had high quality standards and services provided had no leakage.

LESSONS LEARNT

(i) **Creation of organizational knowledge:** Developing housing project under self-financing scheme was a good initiative by DDA and in turn it was a great responsibility. Money was paid in advance and expectation of allottees was safe housing with required quality standards. Besides, proper documentation was necessary to demonstrate the manner in which quality standards were achieved. For this housing project, which was most prestigious project of DDA at that time, proper quality standards were achieved and complete records were kept. Besides, certain initiatives were taken regarding systems for implementation. This was unprecedented. Thus, methodology for implementation evolved for this project is useful for future projects as well.

(ii) **Achieving intended results:** Inputs in building construction are materials and workmanship. Results, as planned, would be achieved only if at each stage of work, these are verified and checked with reference to the standards laid down in the book of specifications and applicable codes. But to know and show that these are getting verified at the designated stage and frequency, a system is required to be put in place. This is what was done for this project, as is evident from the above narrative.

(iii) **Process Improvement:** Adoption of 112½ degree bends for drainage and sewerage, reduced the possibility of leakage to a great extent. Residents were happy and it gave satisfaction to Engineering professionals who worked for the project.

11

STRENGTHENING OF FOUNDATIONS FOR A HOUSING PROJECT AT DELHI

During early eighties, DDA awarded the work of a number of Self Financing Schemes, to different contractors at different places of Delhi. At Kishangarh, in South Delhi, under Self Financing Scheme, two and three bedroom flats were awarded to a contractor. At ground and first floor-three bedroom flats, and at first and second floor-two bedroom flats were constructed. A newspaper report mentioned that houses under construction, at Kishangarh were not having foundation. It was a sensational news. Of course, after inspection, it was found that size of foundations was less than required.

At that time, KB Rajoria was working as Chief Engineer in DDA. The higher management gave him responsibility for attending to deficiencies. To ascertain the exact situation, all the blocks were inspected carefully. It was noted that there was no sign of distress. Therefore, there was no immediate danger about safety of structures. Thereafter, the soil adjacent of foundation was removed in all the blocks, to ascertain exact size of foundations. It was found that out of twelve blocks, only in three blocks, the size of foundation both in depth and width was less than required according to drawings. Therefore, it resulted in more earth pressure on ground, compared to requirements according to safe bearing capacity. Thus, foundations were considered unsafe. For safety and to deal with any adverse situation, it was considered necessary to strengthen the foundation. Therefore, it was decided to evolve such techniques and procedure to strengthen foundation, which could be implemented, without damaging these buildings.

Possible solution to the problem was discussed with a number of eminent structural engineers. Some engineers advised that there was no need to strengthen the foundation, because there was no distress. Besides, the load on foundations was not likely to increase in future. Other engineers mentioned that any excavation around these buildings could make foundations unstable. Therefore, around these building, about 3 meters wide RCC apron could be provided. But these solutions, did not help in strengthening foundations to bear the design load. Therefore, it was decided that foundations should be strengthened. Even otherwise these buildings were to be occupied by individuals who paid for safe houses. Therefore, they should be satisfied about safety.

To start with, subsoil investigations were done. It was found that up to the depth of 5 to 6 meters there was medium density silt and further below hard rock was available.

The safety of existing foundation was analyzed and it was found unsafe. The required factor of safety, according to design Codes, was 2.5 whereas these foundations were having factor of safety between 1.3 to 2.1. To find solutions, Shri Sanjay Gupta, a foundation engineer and a consultant of repute was engaged. He carried out structural analysis. His proposal was that on both sides of foundation, Under-reamed piles should be provided. These piles should be connected at the foundation level by RCC beams. Thus, part load of the building would be shared by Under-reamed piles. This proposal for strengthening of foundations was adopted. Still, on account of the sensitivity of the issue, it was considered desirable to take opinion of top level experts. Those days, Dr. OP Jain, an expert of structural engineering and retired Director of IIT Delhi was residing at Delhi. We discussed our problem and proposed solution with him. His opinion was that proposal for strengthening of foundation was proper and buildings will become safe. It was decided to implement the strengthening proposal given by Shri Sanjay Gupta.

Thereafter, structural designs were worked out and detailed drawings were prepared by Shri Sanjay Gupta. The structural designs were got checked from Prof. M Ahuja of IIT Delhi. This design was based on the detailing done to work out share of load to be taken by existing foundation and under-reamed piles provided on both sides of the wall. These piles were to be connected by RCC beams, to be constructed by making hole in existing brick wall. The space between beam and wall was to be filled by grouting concrete. The spacing between beams was decided on the basis of load bearing capacity of piles, so that additional load could be shared by piles. Besides, along the brick walls, some reinforcement bars were provided so that additional load would be transferred to beams without cracks in brick work. Detailed structural drawings were prepared accordingly and thereafter estimate framed.

After tendering the work was awarded to a contractor. All the walls, which were to be strengthened were identified. The flooring of rooms adjacent to these walls was dismantled. On outer side of the buildings, pavement was dismantled and wherever required water supply lines, sewer lines and electrical supply lines were shifted so that these service lines were not damaged during piling. Thereafter on specified locations, on both sides of the wall, boring for Under reamed piles was done. In these bore holes reinforcement was placed and concreting was done. Piles were cut up to the level of foundation. Between these piles and wall, the space was filled with earth and lean concreting was done up to bottom level of the beam. Opening was made in the wall for the beam. Shuttering was done and then after providing reinforcement, the concreting of beam was done. Plain concrete was also filled in the remaining space of wall, around the beam. After curing of the beam, the space below the beam, inside and outside the room, was filled with earth. Thereafter lean concrete and flooring done, at the level of existing flooring inside the room. On outer side, wherever pavement was dismantled or services were shifted, the work was redone to restore original position. At specified places strain gauge were provided. Load testing was done to check the safety. It was noted that piles were also sharing the load. Thus, it was proved that the required safety factor was available. The work was accordingly completed.

There was one more important aspect. These so called foundation less houses were already allotted and after reading newspaper reports, allottees were unhappy and applied to DDA for cancelation of allotment. These allottees were directed by Housing Department to contact Engineers. Some of these allottees were top level engineers of CPWD and Railways. Total status of project and details of strengthening work done was explained to them. They were satisfied with the work done and agreed not to request for cancelation of allotment. They all occupied these houses. Even after number of years of occupation, no shortcoming came to their notice. It was a matter of satisfaction for Engineers who participated in this strengthening work. In fact all team members deserve appreciation.

LESSONS LEARNT

(i) **Customer Satisfaction:** It was necessary that competent authorities, who assign such sensitive work to a team of Engineers, should be fully confident about the capability of team members. Otherwise, there can be doubts about innovations and job may not be accomplished. Besides, the team should always consult different experts in the area of specialization. Proposals received are required to be examined thoroughly. Moreover, team members should be convinced and confident about evolved solution and methodology. Above all, those for whom the work is undertaken, i.e., the end user, should be satisfied by the work done. Those who were to live in these flats were informed about action taken. They also inspected when strengthening work was in progress.

(ii) **Identifying Innovative Engineering Solution:** Focused attention is required in such cases, like concentrating on the 'eye of the fish' in Mahabharat epic. The team was entrusted with and concentrated on finding a solution rather than finding fault. Thereby there was no diversion of time and energy on activities not in purview of the team. Many times, teams get distracted by likely sensational outcomes but lose their direction and focus from the task at hand.

(iii) **Crisis Management and effective solution to the issue:** Choice and adoption of remedial solution which can be implemented within the boundary conditions can only lead to success. There could be a number of possible solutions to a situation. In the present case buildings were already completed. There was limitation of space (door opening, room size, ceiling height) for any equipment to work from within the room. Thus, use of Under-reamed piles was the appropriate solution which could be carried out without resorting to any dismantling.

OPD BLOCK AT LNJP HOSPITAL COMPLEX DELHI

Loknayak Jaiprakash Narain (LNJP) Hospital and other adjoining hospitals are part of oldest hospital complex of Delhi, situated near Delhi Gate. It is in close proximity to both Old Delhi and New Delhi. Other hospitals in this Complex include Maulana Azad Medical Collage, GB Pant Hospital and Guru Nanak Eye Hospital. The hospital receives patients from not only Delhi but from neighboring States as well. Lord Irwin laid down the foundation stone of this old hospital and it was named after him. In November 1977 the name of Irwin Hospital was changed to LNJP Hospital. Buildings of LNJP hospital were very old compared to buildings of other hospitals in this complex. In late eighties, buildings started showing sign of distress and just to keep buildings in proper shape, time and again requests were received by PWD from hospital authorities for repair and renovations. As required, repairs were done by PWD, but the condition did not improve over time. Detailed inspection was done by PWD Engineers and it was concluded that buildings were beyond economical repair. Therefore, Survey Reports were submitted to the Government for demolition of these buildings. Simultaneously, the planning for developing new complex was started with proposal to dismantle existing buildings and constructing multistoried buildings catering to requirements of increased work load and need for specialty departments, befitting a modern hospital.

The status did not change up to early nineties, even though buildings were getting further deteriorated. At one stage, PWD advised medical authorities to vacate these buildings. But they were helpless as the work of such a busy hospital could not be stopped or closed. The Out Patient Department (OPD) Block was in such a bad shape that a wire net was provided below the ceiling to avoid falling of plaster and parts of ceiling on users. But the big question was, if accident occurs due to failure of the building, what would be the consequences? OPD Block was meant for immediate medical assistance as it accommodated Emergency facilities as well. If accident would have occurred in this Block, disastrous consequences were unimaginable. Warning banners were displayed by PWD at important locations in the complex to make public aware of situation. Minor accidents did occur, but fortunately for the Hospital Authorities and PWD, there was no major accident.

It was a deadlock. Building could not be dismantled as there was no alternative space to run the hospital facilities. Since buildings could not be vacated, it was not possible to dismantle. In turn, the development work of multistoried complex could not start. Medical authorities and PWD Engineers discussed at top level to find workable solution. To break this deadlock Dr. Harshvardhan, the then Hon'ble Health Minister Government of Delhi, was requested to visit the hospital complex. All the affected locations were visited, jointly with team of doctors headed by Dr. Bharat Singh, Medical Superintendent and PWD team headed by K B Rajoria as Chief Engineer along with Shri AK Garg, Superintending Engineer (Civil), Shri SC Khurana, Superintending Engineer (Electrical) and other PWD Engineers. Thereafter, a meeting was conveyed by Hon'ble Minister in the Chamber of Medical Superintendent. Different alternatives for shifting Emergency services of OPD Block to adjoining hospital complexes were considered but not found feasible. It was finally decided to construct a OPD Block and a Ward Block in those locations which were not near to existing buildings and where no multistory building was to be constructed in future. Dr. Harshvardhan desired that to break the deadlock and avoid accidents in hospital complex, these two buildings should be constructed in minimum possible time. After discussion, as desired by Hon'ble Minister, PWD agreed to construct both buildings in four months period provided drawings are drafted immediately, and approved by the Medical Superintendent. Besides, financial approval to be granted by Government in next two days. Hon'ble Minister accepted proposal given by PWD and action started on war footing.

It was decided that Architect and Engineers of PWD should work in the office of Medical Superintendent to prepare drawings. Shri DK Nigam, Architect PWD, was sent to the office of Medical Superintendent with drafting aids. He was assigned the task of preparing drawings and get approval of Medical Superintendent. It was a tall order. Dr. Bharat Singh Medical Superintendent was a dedicated and sincere officer. Shri Nigam could finalize drawings both for the OPD Block, a single storied air-conditioned building and a Ward Block, a two storied building with lift and other facilities. Dr. Bharat Singh approved these drawings on the same day. Immediately thereafter an Estimate was prepared by PWD Engineers and sent to Finance department for approval. Hon'ble Finance Minister Shri Jagdish Mukhi was informed on phone and he appreciated the initiative taken by Dr. Harshvardhan. He very kindly promised to grant approval without delay. The needful was done by the Finance Department in two days period and Financial Approval conveyed to PWD.

Thus, PWD had preliminary drawings for two buildings and the financial approval of the Government, to proceed with the project and complete in four months period. The planning team was headed by Shri SS Mandal Superintending Surveyor of Works, and Architect team was headed by Shri AK Pathak, Senior Architect assisted by Shri DK Nigam Architect, Shri AK Garg Superintending Engineer (Civil) and Shri SC Khurana, Superintending Engineer (Electrical), were heading implementation teams. A programme of action was worked out in the form of Bar Chart. Besides, both Superintending Engineers were advised to select their teams for implementation. They did so and officers selected for implementation of project from civil wing included Shri Vivek Bansal EE Shri Manoj Kumar, Shri Daljeet Singh AEs and Shri MA Ali & Shri RL Srivastava JEs. From electrical wing, Shri PC Jain, EE Shri Sarang, and Shri Sharma AEs and Shri Tingle, JE, were selected.

Important steps taken to implement the project expeditiously were as follows.

Detailed drawings were drafted in two days period. The Estimate for work with detailed BOQ, up to plinth level, prepared and Work Order given to a reliable contractor by negotiations across table. Thus, work of excavation, base concrete, masonry up to plinth level and DPC was started immediately and completed by the time remaining work awarded after call of tenders. The decision to split the work up to plinth level in one part and remaining second part was taken to reduce the time for completion by about one month.

For both buildings, detailed drawings including services and structural drawings drafted in one week time.

For each of the two buildings separate tenders were invited. Tenders received were scrutinized on top priority. For Ward Block, the quoted rate by the Contractor was found reasonable and work was awarded immediately. Rates given by all the contractors for the OPD Block were very high and it looked a pooled tender. Still the contractor who gave lowest rate was called for negotiation but he did not reduce rates to reasonable level. So tenders were recalled with reduced time for completion. The rates received on second call were reasonable and work was awarded. While awarding work, the contractors were told that in case of any difficulty, they should meet KB Rajoria the Chief Engineer and explain their requirements. They were assured that difficulties would be solved then and there. Of course, no delay would be tolerated. Contractor gave commitment to complete the work within time specified.

For electrical wing works were awarded for electrification, air conditioning and lift separately, with commitment from contractors to complete their work in specified time.

Detailed bar charts were prepared for both the projects, covering even smallest activities. The target dates for different stages of work were worked out. The bar chart included details for all activities including internal electrification, lift, air-conditioning etc. All officers and contractors worked very hard practically day and night. The programme was reviewed at regular intervals and there was no slippage. The contractor for the OPD Block came with some difficulties to KB Rajoria, Chief Engineer. As committed to him, difficulties were solved without taking any time. The construction site was visited on frequently to review the progress and encourage the site staff as well contractors. According to commitment to Dr. Harshvardhan, Hon'ble Health Minister, both the buildings were completed within four month and handed over. It was an unparallel achievement.

It is an example of giving commitment and fulfilling the same. All the wings of PWD worked hard and gave whole hearted cooperation to achieve the objective. The team work spirit is worth appreciation, Hats off to team members. A challenging and difficult task could be achieved. It is needless to mention that completion of this small project, costing about Rs 6 Crores, raised moral of PWD Architects and Engineers. This impossible could be done on account of political commitment at highest level as also cooperation and coordination between Delhi Government, Medical Department and PWD. Grateful to political leadership, i.e., Dr. Harshwardhan the then Health Minister, and Shri Jagdish Mukhi the then Finance Minister, Government of Delhi to have confidence on commitments given by PWD which were duly fulfilled.

After these buildings were occupied, the dismantling of existing buildings started. It was breaking of deadlock and beginning of development of new multistoried complex.

LESSONS LEARNT

(i) **Leadership and Motivation:** As part of the strategy, support of the top level in Delhi Govt was solicited by requesting the Hon'ble Health Minister to pay a visit to the hospital and chair a meeting with all concerned. This ensured speedy administrative and financial approvals from the Govt (essential for taking up of work by PWD) as well as quick approvals of plans and specifications from the Hospital authorities. The approval was given with a rider that the new building must be completed within four months' time. PWD had done some homework earlier also and the same was agreed to with the firm belief that it would be done.

(ii) **Team Spirit for achieving objective:** Teams were constituted from Civil, Electrical and Architectural wings of the Department for this time bound task. The top brass of PWD gave team leaders an opportunity to select their own team members. By this process the chosen persons also felt a sense of commitment and pride. The team leaders also had confidence and commitment of achieving the target.

(iii) Frequent visit to the work site by the Chief Engineer not only resolved day to day issues but also developed a comradeship with the field staff resulting in developing bonding with the work. Also, this strengthened the belief among the staff that the target would be achieved.

(iv) **Operational Planning and Control :** Important decision was splitting up of the Civil works into three parts and assigning it to separate agencies. Thus, creating room for deployment of adequate resources. This also saved at least a months' time. At the same time prescribed procedures were followed. Thus, the principal of dividing the problem (or solution) in parts worked very well here.

(v) **Time Management:** During execution, there was efficient and effective coordination between department and contractors. Project could be completed in record time of four months. It is an example of human desire and courage for accomplishment. Where there is a will, there is a way.

During the year 2000 the Secretariat Building for Govt. of Delhi was completed and since then it is occupied by Government of Delhi for offices of Chief Minister, Council of Ministers, Secretaries to Government and the concerned staff. It is an impressive building adding to landscape of Delhi. It has an interesting background, as brought out here.

Hostel Building for Players – During 1982, Asian games were held at New Delhi. For this event various countries of Asia including Japan, China, Indonesia, Pakistan etc. participated. Different games and sports were organized for which numbers of facilities were developed by different Government Agencies. The Indoor Stadium Complex was developed by Delhi Development Authority. In this complex, besides several other structures, one ten storeyed building was constructed. This building was named as Player's Building and meant for the hostel for players during Asiad 82. Because of change in priorities, after completion of framed structure, the work on this building was stopped. Therefore, after Asiad 82, this building remained in incomplete shape. Authorities wanted to use this building for an alternative purpose and several proposals were considered. During 1988, this building was handed over to Indraprastha Health Corporation Limited, to develop it for a hospital. They started repair and remodeling of the building. On account of differences between Government and Indraprastha Health Corporation, this work did not proceed and stopped in incomplete shape. During 1995, PWD Delhi was given responsibility to remodel and redevelop this building as Secretariat. At that time, Shri KB Rajoria was Engineer-in-Chief, PWD, Shri IM Singh was Chief Engineer, Shri DS Sachdev Superintending Engineer(Civil) and Shri Anil Puri Superintending Engineer (Electrical).

Building condition assessment – as already brought out, During 1981-82, only ten storeyed framed structure was constructed and building was left incomplete. Thereafter, Indraprastha Health Corporation did some addition and alteration. The building was having three wings in star shape. The total built-up area was 40,000 sqm and area at ground floor was 9000 sqm. There were 1600 floors columns, 3200 beams and 1200 slab panels. It was planned to have 400 hostel rooms for players with attached toilets. There were fifteen terraces at different levels, where planters were provided. During 1987-88, the remodeling was done for conversion of building as hospital. Besides, at ninth floor, four bays were added in all the three wings. The building was in a bad shape at the stage of taking over by PWD during 1988-89. It was observed that the reinforcement was exposed at several places. There were cracks in the concrete and at places some concrete had chipped off. Isolated differential settlement was also observed in ground floor walls. There was water logging. Fortunately, no structural cracks were noticed. It was decided to do retrofitting of the building to make it structurally safe and sound.

Study of distress pattern and non-destructive testing of RCC work – Computer generated model studies were carried out to find out the distress pattern of the building. It was noted that some RCC members were severely distressed. By and large, members on outer side were more distressed compared to other members. Beams near expansion joints and sunken floors were also more distressed. There were cracks in concrete on account of rusting and spalling of reinforcement bars. Apparently sunken areas were more distressed due to collection and seepage of water in slabs. Non destructive testing was carried out at different places in such a manner that all critical locations were covered. (a) By Rebound Hammer test, the compressive strength of concrete was ascertained at different locations, (b) By UPSV Test, Pulse Velocity was ascertained at different places. It is measured in km/sec and is indication of quality of concrete. Correlation with quality being (i) Excellent- above 4.5 (ii) Good- 4.5 to 3.5 (iii) Medium up to 3 and (iv) doubtful below 3. (c) Half Cell Potential test was done to get information regarding risk of corrosion of reinforcement bars embedded in the concrete. This test was undertaken at places where visible cracks or rust stains were absent. It was done as per ASTM-C 876-1991. The range of test value indicated chances of corrosion in percentage. (i) Value- 350 and above - chance of corrosion was 90% (ii) Value- 350 to 200 - chance of corrosion was 50% (iii) Value -200 and less - chance of corrosion was 10%. (d) Allowable limit for chloride was 0.25% and allowable limit of sulphate was 0.39% of mass of concrete. Tests indicated that chloride and sulphate were within the limits. So, there was no worry on account by chlorides and sulphates. The distress report was documented in a systematic manner. All structural members such as beams, columns and slabs were individually examined and distress details recorded. This included honeycombing, corrosion of reinforcement, laboratory and non destructive test results etc. On the basis of this report, decision was taken for each member, (a) whether it was to be dismantled, (b) whether it required major repair, (c) whether it required patch repair and (d) whether it required no repair.

Repair and Rehabilitation of Building – After detailed investigations, the strategy for repair and rehabilitation work was drawn. Depending on condition of individual member, the repair/rehabilitation work was done. In first stage, the chipping of un-sound concrete was done. Thereafter, the reinforcement was cleaned with wire brush, emery paper and sand blasting. On reinforcement bars, anti rust agent (Nitro Zink Primer) was applied. On concrete binding agent was applied. For columns and beams, binding agent applied was Nitro bond-EP and for slabs it was Nitro Bond-AR. Fins of columns were knocked down and then restored by using micro concrete. In beams and columns if damage was more than 60% then these were dismantled and re-cast. For less damaged members, only localized concreting was done. Wherever reinforcement was found excessively rusted, additional reinforcement was added. For isolated repair work, groove was cut along crack line and grout was injected with pump. To fill finer cracks, epoxy grout was injected. In excessive honey comb area, richer mix was used for the grout. To check quality of work, several tests were performed on the repaired work/retrofitted work, which included cube test, polymer content test, USPV anti-rust test etc. For restoration work, approximately Rs. 150 lacs were spent. This amount was only about 10% of cost of structure if reconstructed afresh at that time.

Architectural Planning – The building was remodeled as a Secretariat for Delhi Government. For Architectural planning, Shri Raja Adheri Consultant Pvt. Ltd, a famous Architect from Mumbai was selected. For implementation of project selection was by shortlisting contractors. The work was awarded to M/S Ahluwalia Contractors (P) Ltd. The original frame work of RCC structure was not changed by the Architect for architectural detailing. Additions and alterations, as considered necessary were also incorporated. The broad planning of the building was done in following manner:

Ground floor – At this level VIP Parking, Auditorium with 200 seats, Banquet Hall, Canteen and commercial facilities such as Bank, Post Office, shops etc. were planned. Besides, scooter parking was also accommodated in this floor. Air conditioning plant and electrical generators were located in Service block. One multi-purpose hall was also planned at this floor; (b) First floor- At this level, two conference rooms and office space were planned; (c) Second floor- Chief Minister's office and the other office space was planned; (d) Third to eighth floor- open office space on modular concept was planned; (e) Ninth floor- the guest rooms were planned at this floor.

Architectural Planning – Interior spaces were planned according to requirements of Secretariat Building. The building was made fully air conditioned. Besides, for control of entry to the building, CCTV cameras were installed. Moreover, exterior elevations were remodeled and permanent finish was provided, befitting to an important Govt. building. It was planned as an intelligent building with harmonious amalgamation of energy management, fire detection system, security surveillance data, communication network, audio & video and tele conferencing facilities. The project was completed in a little over 2 year period. The Delhi Secretariat is functioning since then in this building.

LESSONS LEARNT

This Building was left incomplete as only structured work was done by the agency. After a few years, it was given to another agency and some additions/alterations were done and was not completed. The building was thereafter handed over to PWD Delhi for making it suitable for Delhi Secretariat. It was a big responsibility to do retrofitting and undertake addition/alterations to make it suitable for the office complex. Undertaking structural testing and thereafter repair/rehabilitation work involved risk. It was the done by PWD Engineers in a systematic manner. The Successful completion of repair and rehabilitation work speaks of high order of confidence of Engineers who overtook this task. The alternative was to demolish the building structure, which was avoided thus, there was saving for the Government. The courage and technical competence of engineers is worth appreciation.

Secretariat Building for Govt. of Delhi -
Entrance gate and front elevation

Secretariat Building for Govt. of Delhi -
View with garden

Secretariat Building for Govt. of Delhi -
View of garden

Secretariat Building for Govt. of Delhi -
Side Elevation

1. BACKGROUND AND APPROVAL OF THE PROJECT

1.1 During early twentieth century, Parliament Building was constructed as a part of Capital Complex. After independence of India, the activities at Parliament Building multiplied many folds. Difficulty was felt on account of non availability of adequate space for storage of books and documents and running the office of Parliament Secretariat. During early seventies, it was decided to construct Parliament Secretariat Extension Building at No 1 Parliament Street, adjacent to the Parliament complex. This building was completed during 1975 and the Secretariat was shifted from Parliament house to this building. The difficulty for storage of books and publications continued. Several alternative proposals were examined by Parliament Secretariat, in consultation with Central PWD. Finally, during a meeting held during August 1988, under the Chairmanship of Hon'ble Prime Minister of India, it was decided that Parliament Library Building should be constructed within Parliament Complex, adjacent to Parliament Secretariat Extension Building. The Ministry of Urban Development and in turn CPWD was asked to go ahead with the project. After meetings at high level, several alternatives were considered for proper plans of the proposed library building. It was finally decided that an eminent Architect from Delhi should be selected after architectural competition to design the building. The committee headed by Hon. Minister of Urban Development, selected M/S Raj Rewal Associates as Architects. Thereafter, CPWD was directed to go ahead with the project. Accordingly preliminary drawings were prepared by consultants and approved by competent authority.

1.2 The Architectural form of the building was conceptualized with an aim to achieve a low-key architectural expression signifying sagacity and spiritual elegance rather than to compete with the power of the Parliament House. The square library building sits on a triangular plot with the planning ideals of the temple of Raunakpur and Datia, complementing the circular Parliament House. The height of the building has been restricted to the podium level of the Parliament House. Only architectural features of the glazed crystalline forms of 'domes' protrude above the podium level, keeping the height well below the roof of the Parliament House. The building has four main entries. One for the VIPs, others for Members of the Parliament, Scholars and the Public. In addition, separate entries provided for receiving bulk supply of publications, and kitchen supplies. Space was demarcated as Reception, Auditorium, Museum, Audio Visual, Bureau of Parliamentary Studies and Training, Library, The Library Reference, Research, Documentation and Information Service Museum, Archives, Committee Rooms as also MPs Reading Spaces. Public spaces and administrative spaces areas were located at the ground and the first floor. Basement was meant for stacking of books, storage and service requirements. Red stone cladding provided on the external surfaces, matching the finish of the Parliament House. Internally the library building finished with a variety and combination of materials. Total area of 60,460 sqm including parking for 212 cars in three levels was provided. The area is nearly one and half times that of the Parliament House Building. The sanctioned cost of the project was Rs.192 Crore. The Lok Sabha Secretariat approved the architectural plans in February 1993. Thereafter approval from various local bodies was taken.This included NDMC, DDA, Central Vista Committee, Delhi Fire Service, and DUAC. Bhoomi Poojan was done by the Hon'ble Speaker, Lok Sabha on 17th April 1994 and the work on project started.

2. EXCAVATION AND FOUNDATIONS

2.1 **Excavation**- the Parliament Library building consisted of two floors above ground level and two basement floors. The contiguous car parking area had three levels. This involved huge excavation. Detailed soil investigation revealed presence of soft to medium strength rock from 1.2 m to 13.5 m below existing ground level. It also brought out that the underlying soil was sandy silt to sandy clay. Monitoring for any uplift or settlement by installing gauges around Parliament House Building was done. The foundation system comprised raft foundation covering about 19,000 sqm area. It supported 855 columns. 1752 vertical rock anchors provided to counter uplift pressure due to the buoyancy. The structural design of the raft foundation was done using 'Finite Method' of analysis of the heaviest loaded area. The raft was laid in a slope of 1 in 150 for drainage purpose. Diaphragm wall was provided so that about 9 m deep excavation could be carried out in limited space available. Inclined and vertical rock anchors to take horizontal and vertical forces were used. Suitable waterproofing treatment was done to keep the entire basement area damp proof.

2.2 **Foundation**- Diaphragm wall of 60 cm thickness in panels of 5 meter width was provided all around the excavation area to avoid any possible disturbance to soil strata supporting the Parliament House Building. Also, there was space constraint to permit deep excavation by conventional methods. The bottom of the diaphragm wall was anchored upto a depth of 5 meters into the rock. Two inclined rock anchors of 60 Tonne capacity at each level were provided in each such panel for the stability of the diaphragm wall during execution. The number of levels varied from 1 to 3 depending upon depth of the wall. Structural design was generally carried out in accordance with IS: 9566-1980. Minimum 0.2 % steel was provided. For detailing, international standards were consulted. Work of diaphragm wall was started with construction of Guide wall around. Once the rock was met, chisel of about 2 Tonne weight was used to break the rock. During the entire operation of excavation and construction the excavated trench was kept filled with Sodium based bentonite slurry of 5% concentration. This had the property of stabilizing the sides and prevented them from collapsing. Once the excavation was done upto the desired level, the reinforcement cage was lowered. Trench was flushed thoroughly before the cage was lowered. Two gaps were left in the cage for inserting tremie pipes used for under water concreting of the diaphragm wall. Trumpet Pipes at 1.75m, 4.25m, and 6.5 m below the cut-off level were left at the time of preparation of the cage with necessary pockets to accommodate the bearing of the anchorage system. A mesh of 12 mm diameter steel reinforcement was provided around the trumpet pipe on the soil side of the cage to bear the shear stresses. Tremie pipe, 20 cm diameter was used in required length by connecting several pipes. The pipe was gradually withdrawn as the concreting progressed, taking care that lower end is always embedded in concrete. Gradual pulling out the pipe during concreting created surges which facilitated flow of concrete and compaction. The diaphragm wall panels were tied together by continuous capping beam above the cut off level. In each cage, 3 GI pipes 65 mm diameter were placed vertically for grouting operation.

2.3 **Concreting of foundations**- Concreting of a panel was done in one operation to achieve a joint free continuous surface. The next panel was cast by the side of another completed panel so as to form a good joint and continuous leak proof diaphragm wall. Joints between two successive panels were made by using a stop end pipe, as side shuttering, of diameter 58.5 cm, of 1cm wall thickness. Before lowering the cage for a successive panel, the circular stop end pipe of the previous panel was pulled out by deploying hydraulic jacks after the concrete had partially hardened leaving a semi-circular key in the preceding panel which was occupied the male end of the succeeding panel to form a perfect joint. A total of 84 panels in 410 m of periphery were executed in less than 4 months' time. These panels were anchored by 360 number inclined rock anchors. About 1.2 lakh cum of soil and 0.95 lakh cum of rock were excavated without blasting, partly under water. Heavy machinery was used to carry out such huge excavation. The site being a security area, movement of trucks was restricted to night time only. Curtain grouting was carried out, as recommended in soil investigation report, to arrest water inflow. This reduced pumping to a great extent. Combination of heavy equipment like hydraulic excavator, pneumatic pavement breakers, rotary drills, hydraulic rock splitter, hydraulic hammer were used to carry out excavation in rock strata.

3 SUPERSTRUCTURE

3.1 **Dome structure above the roof level** - the structure has twelve domes of special geometry, with various shapes and sizes. Some of these domes are opaque and some partly opaque and glazed. The central focal dome is fully glazed to provide diffused light into the basement level. The crystalline form of glass and steel present an attractive feature during the day and night. The diameter of these domes is varying from 14 meter to 35 meter, having varying geometry. It is an important architectural feature of this building. The structural part for the dome consists of tubular truss shaped to form skeletal dome roof arranged in hexagonal/octagonal and square pattern and supported through ring beam and reinforced neoprene bearings on RCC columns for permitting movement due to temperature variation. The octagonal or hexagonal pattern formed by the steel frame are filled by placing precast concrete Bubbles. Except for the dome over the auditorium which has false ceiling, all others are visible from inside. The structural work for the auditorium is in steel. For remaining areas, it is either in stainless steel or carbon steel matching with interior décor. The external cladding on these domes is varying and is supported on space frame. The cladding consists of 'high performance fibre reinforced concrete' Bubbles (HPFRC) of M-50 grade, sand stone, granite, glass sheet and glass blocks. HPFRC as one of the cladding materials was selected according to the advice of Structural Engineering Research Centre, Chennai. The surface of the domes is spherical or cylindrical in shape. A three-dimensional model in Auto CAD was prepared for each dome to get the coordinates of joints. Optimum size of the structural steel tubular members was fixed by trial-and-error method. At each junction, 6 to 12 members of the supporting space frame were meeting. As such, welded joints were ruled out on accounts of likely distortion in shape and in turn appearance. M/s RFR, Paris, who were the consultants, carried out detailed engineering of domes which consisted of joint detailing, connection details between the steel Ring Beam and the dome structure, connection between Bubbles with steel dome structure and support system of the steel ring beam on existing RCC columns. Cast steel joints were recommended. This required precise fabrication. Sensitive to geometrics, deflection was also checked in accordance with IS:800-1984, limiting deflection within 1/325 of the span.

3.2 **Panelling** - Horizontal reinforced precast glass block panels were provided as skylight in the cut outs of precast fibre reinforced concrete bubbles of library and museum domes to allow diffused light. The glass bubbles are frosted from inside. Glass blocks are of size 190x190x80 mm supported on 25 mm wide RCC ribs in both the directions. The unit is made up of 6 mm heat strengthened and heat reflective glass as the outer surface along with three layers of heat strengthened 10 mm thick float glass with layers of resin between two layers of glass. Two layers of glass, bonded by resin, are separated by hollow Aluminium spacers to form a gap for insulation purposes. Heat insulated and laminated double glazed safety glass units provided, in combination of the glass blocks walls, along the periphery of domes.

3.3 **Water proofing** - The precast HPFR concrete bubbles do not have any structural connection to each other, as these are simply placed on steel structure with a gap of 20 mm all around. This gap is provided so that full contraction and expansion could take place due to change in weather. The length of joint between bubbles work out to about 0.5 meter for every square meter of the surface and thus it was necessary to have a water proofing system which could permit expansion / contraction and highly reliable in performance. The joint was first sealed with Silicon sealant. A coloured primer was used to avoid any human error. The water proofing, in a width of 250 mm across the joint, was carried out in three layers using Polycoat Products, short listed for the purpose. This was provided and executed by M/s JBM Engineers Pvt Ltd. A membrane of minimum thickness of 1.5 mm was achieved in 5 coats. This was protected by 15 mm thick layer of cement lime mortar in two layers [1:1:2:2, (Cement, Lime, Surkhi, Sand)] with polypropylene fibre. A prototype was tested under water head of 15 cm over a treated joint for 24 hours and observing the bottom for any signs of dampness/ leakage.

3.4 **Stone work** - Different kinds of stone work was done for cladding to conform to the external façade of the Parliament House. Silicone based water repellent paint was applied on all the exposed external surface to minimize weathering effect. The Red Sand stone used in most of the locations. This stone was fine chisel hand dressed. The external walls were with quarry finish. Dholpur (white) stone was used for wall lining, copings, rain water spouts, and making jalis. Pink sand stone was used as bond stone in wall lining in combination with white sand stone. About 25,000 pieces of circular cladding quadrants were used as cladding to circular surfaces. To meet with shortage of highly skilled craftsman for such skilled work, and reduce time of such operation, improvised lathe machines were specially used for finishing exposed circular surfaces.

3.5 **Doors and windows** - The door frames, door shutters and architraves in building were made with first class Burma teak wood. Same wood has been used for hand rails, false ceiling and wall lining. Cedar wood and Shisham wood has been used for various type of decorative false ceilings. In order to ensure proper seasoning of timber, certificate from the kiln seasoning plant was checked in addition to usual prescribed tests. Powder coated aluminium sections were preferred for windows, glazed entrance doors and openings in the court yards. Hermitically sealed double glass units comprising two sheets of float glass panes, separated by hollow anodised aluminium spacer used for glazed windows. Since no IS Code was available for float glass at the time of framing specifications, Japanese code JIS R 3202 was referred for acceptance criteria.

3.6 **Flooring, cladding and false ceiling** - Depending upon location, flooring with different material and pattern has been provided in 57,529 sqm area. Function of areas are varying from location to location like stack area for books, reading areas, banquet hall, auditorium, museum, service area, parking etc. Variety of materials like marble, stone, ceramic tiles, Burma teak wood, Rangoli pattern have been used as provided in architectural drawings. Depending upon importance of location, type of wall lining/cladding and use of the area, different types of intricate and elegant type of false ceiling have been provided. In most of the areas false ceiling has been provided with acoustic backing to keep the noise level low. Materials used are Gypsum board, Burma teak wood, stainless steel plank, perforated aluminium plank, solid sand stone and sand stone jali. The supporting frame for the false ceiling has been carefully selected and has been designed for fire safety and durability. Provision for light fittings, PA system and air conditioning grill have also been made to match the intricate pattern of the ceiling as in architectural drawings. Problems associated with making openings in the false ceiling for light fittings and AC grills were solved with local ingenuity, keeping in mind the requirements during maintenance also. Wall linings are provided on functional requirements. Some of the materials used for internal wall lining/cladding include Burma teak wood, 18 mm thick marble slats, slotted marble slats, sand stone jali, granite and tiles. Acoustic backing has been provided in specified areas.

3.7 **Landscaping** - Three distinct areas for landscaping were covered in the project. Selection of plants was based on aesthetic considerations relating to formal and informal character of the open spaces as well as climatic conditions. Existing trees in the open areas of the compound were retained and supplemented with suitable species unifying various areas. The three courtyards, though similar in scale, were designed differently to align with the functions in the adjoining buildings. The terrace garden was landscaped with strategically located planters which presented an appearance of a formal garden and a fore-court for the Parliament House.

4. INTERNAL AND EXTERNAL SERVICES

4.1 **Water supply for toilets and drinking**- Keeping in view the requirement of various areas, nine toilet blocks were provided in the building. Six terminate at the first basement level and remaining three at second basement level. Out of these five toilet blocks were situated along periphery of the building and remaining in the core. Water supply and waste water disposal was planned accordingly. Required 1.5 lakh litres of filtered water was supplied by NDMC through 80 mm diameter Tee connection on Pandit Pant Marg. The water was taken to the underground sump through 100 mm diameter GI pipe. The underground sump had three compartments. First compartment had the mandatory capacity of 1 lakh litre for the firefighting requirement. Once this tank is filled, water spills over the second compartment meant for drinking purposes having full one day supply capacity. The third compartment is for air conditioning purposes and water is filled by tube wells. Water to individual block was supplied through overhead RCC water storage tanks located on every block, top finished with red sand stone to match the walk-way of the terrace garden. Individual tanks were provided as there was no space and there was height restriction for a single master tank on the terrace. Capacity of these tanks was for pumping water twice a day. Hardy-Cross method was used to calculate the head loss in various branches, in order to ensure equal head loss at delivery end of every tank, so that each tank get same residual head. The distribution line was made in two loops. Inner loop was around the focal dome and outer loop along periphery of terrace. Rising main was of 100 mm GI pipe through the shaft between the service and the auditorium block. The entire distribution system was divided into four segments by providing sluice valves to ensure segregation of any particular segment for repair or replacement. Two number 65 mm diameter GI pipe carry water from the water tanks, one serving the first floor as also first basement, and other serving the ground floor as also second basement. Pipes were having reduced sizes according to design. Scour pipe 80 mm diameter were connected to the peripheral drain in the second basement for cleaning of water tanks.

4.2 **Requirement of water for air - conditioning purposes** was 4.5 lakh litres per day. As NDMC expressed its inability to provide filtered water for this purpose, the requirement was met through 2 number of dedicated tube wells installed by NDMC outside the complex, as a Deposit work. Water from the two tube wells being stored in underground water tank of 4.5 lakh litre capacity.

4.3 Keeping the drinking water situation of the city in view, filtered water for horticulture purposes was not made available by NDMC. Use of unfiltered water was ruled out as sometimes it has foul smell. Its use was required for greenery in the indoor courtyards as well. Also, water sprinkler system was likely to get choked due use of unfiltered water. Therefore, three tube wells were developed within the campus towards meeting requirement of water for horticultural purposes, estimated as 2.5 lakh litres during peak summer. These tube wells connected to 80 mm diameter PVC pipe grid laid along the outer periphery of the external area from where branch lines of suitable diameter, according to design, were laid. Each tube well provided with online filter. Pressure relief valves provided near each tube well to cater for a contingency when all the valves are closed. Pop up sprinklers provided for irrigation of lawns. Drip sprinklers provided for flower beds.

4.4 Disposal of waste water was planned, so that waste water from the five peripheral toilet blocks at ground and first floor flowed directly into manholes outside the building by gravity. For the four toilet blocks located in the core of the building and for the toilets in the first and second basement, the sewage is collected in five sumps located in the second basement. These sumps were designed for 24-hour retention plus 25 percent for dead storage. The horizontal pipes are 150 mm cast iron, laid in minimum slope of 1 in 200. Location of bends was such that all were easily accessible for cleaning and replacement if required. Each sump was provided with two pumps, one normal, other standby, operating automatically discharging to external manholes at ground level and then to the sewer line leading to the municipal manhole. For connection to the Municipal trunk sewer line, it was found that a manhole with deeper invert level was available near Patel Chowk. NDMC laid a sewer line from Patel Chowk to the manhole on Talkatora road near Parliament Library building, as a deposit work. A sewer line within the building campus, connecting all the manholes and having a length of about 400 m was laid and connected to the Municipal manhole for disposal.

4.5 It was necessary to provide drainage of the raft or the second basement to cater for the likely water from testing/operation of the sprinkler system, scour pipe discharge after cleaning of the overhead water tanks, leakage, if any, from the wet risers, sewerage system and seepage, if any, from the basement walls and floor due to water table rising almost up to ground level. A peripheral drain with perforated cover along the internal and external periphery provided with minimum slope of 1 in 100, having minimum depth of 150 mm and maximum depth of 500 mm. Eighteen number of drainage sumps were provided each with 4 HP automatic electric pump, discharging into the external trunk storm water drain.

4.6 The rain water in the internal courtyards was guided through drains, having perforated covers, along the periphery and drained to the external storm water drain, connected by pipes through the RCC walls provided at the casting stage. Drainage of roof was achieved using rain water vertical gutters, drainage spouts and water falling over chajjas and then to the lawns below. The huge terrace garden provided with greenery generated dead leaves etc. Slotted strainer pipes were provided to prevent choking of the drainage spouts and the pipes. Drainage of external areas were provided through RCC storm water lines of diameter 300 to 450 mm, laid along the periphery of the building to avoid stagnation of water along the walls. These were connected to gully gratings to collect water from different areas. The storm water mains then connected to the trunk storm water main passing through the campus.

4.7 **Power supply** - The design load of library building was about 7 MW and maximum demand was 5 MW. The power supply was given by NDMC by installing a 33kv Sub-station. The power demand of 5 MW was met from the power supply received from this Sub-station and another 33 kv Sub-station near National Archives. Besides Parliament House sub-station was also interconnected to library Substation. It was located in basement at 5.4 m below ground level. The main load was for air-conditioning and winter heating system. It was designed with maximum emphasis for minimum vulnerability to fire. This was the main sub-station in which dry type of transformers were used. The capacity of sub-station was 1250 KVA. MT vacuum type breakers were provided. The HT panel consisted of 15 No of 630 Amps circuit breakers with service of tripping system. LT panels were provided for power distribution system.

4.8 **Stand-by Power supply system** - The building was having two basements and two floors above ground level. It was multi-dimensional building having important areas like computer center, Audio and Video Museum, Audio Video Presentation Center etc. Complete building was air conditioned and no alternative method of air ventilation provided. Any interruption in power supply would result in complete stoppage of activities inside the building and the basement. Therefore, for emergency power supply, two DG sets of 1000 KVA capacity were provided. DG sets for library building were installed in the basement at 5.4 m below ground level. DG sets were very close to auditorium and many noise producing equipment were installed in service area, e.g., Ventilation blower, AC equipment etc. Therefore, acoustic enclosures were provided for DG sets.

4.9 For internal electrical installations compact fluorescent lamps were used. Consulting Architect gave shape, size and type of lamps for each area. The number of fittings and wattage of lamps were designed on the basis of illumination level required in each area, glare, colour scheme of room and light distribution in the area. Illumination level was calculated on point to point basis in an area of 0.5m x 0.5 m or 1.0m x 1.00 m. Committee rooms required different levels of illumination, e.g., (i) 70 lux during multimodal presentation, (ii) 300 lux during normal conference and (iii) 450 lux when video is recording was required. It was decided to provide eighteen down light fittings, each with two CFL lamps of 18 watts. Lighting levels were kept as per 1S-3646 (Part-1) 1992. Other issues considered for lighting design were (i) light controls (ii) Use of copper cables (iii) MCB Distribution Boards (iv) colour coding (v) Energy saving by use of Electronic Ballast. For other areas, the lighting layout, type of fittings, type of lamp and wattage of lamps were decided on the basis of lux level calculations for all areas, other than committee room.

4.10 **Compound lighting system** - Three court yards of the complex were designed with themes of liberty, justice and equality. Besides, garden at terrace was also designed for required lighting. Low level light fittings were provided in these areas.

4.11 **Floor trucking** - In this building there are large halls for reading area, work stations, computer centre, information centre, committee and conference rooms which have no carpets on the floor. So, there was problem of taking wiring for telephone, computers, call bells, table lamps etc on to the furniture such as office tables, work station etc. In order to avoid loose wires on the floor, provision of floor trucking was done in these areas. The floor junction boxes were provided at regular intervals for facilitating laying and maintenance of wiring system. The junction boxes were covered with stainless steel plates. For floor trucking, furniture and work stations have in built provision for wiring path, outlet boxes switches, sockets and switch plates. The socket outlets were provided for office tables and workstations. Cables brought up vertically upto furniture from junction boxes through stainless steel tubes (named as wire managers) and terminated in the proper outlet point or connecters provided on the furniture items. During execution there were a few learning lessons which could be useful for future projects: (a) Location of floor junction boxes were decided prior to decision on location and layout of furniture. There were difficulties in making even minor changes in their location by the user; (b) Floor trucking was terminated in important service areas like AHU room, EDB shaft. These boxes had to be sealed subsequently to prevent flow of condensation water from AHU rooms to EDB shafts and; (c) Great difficulty was faced in providing stainless steel cover plates on pre fixed floor junction boxes especially in Kota stone flooring areas with mosaic strips. The size of the stainless steel plates had to be finally changed to include width of strips.

4.12 **Air conditioning and cold storage system** - The control over temperature and humidity was the basic requirement of functional library of National importance. Therefore, complete library building was provided with air conditioning facility. Non functional areas like service area and car parking were provided with ventilation system. For AC plant room space was provided at 5.4 m below ground level. The location for cooling tower were provided. The design of Chiller was adopted based on recommendation of experts on technical requirements for air conditioning. Other components included cooling towers, AC plant for rear block section and cooler storage.

4.13 **Ventilation system** - Total floor area of Parliament library was 55,000 square meters out of which nearly 45,000 sqm was air conditioned. For remaining area ventilation system was provided. For car parking area equipment were provided for supply and exhaust of air for three basements. For toilets ventilation was provided by cool air of air conditioning system. Ventilation plant for book binding area and conservation laboratory was isolated form air-conditioning system. Also, separate ventilation for kitchen was provided.

4.14 **Fire detection and water based fire fighting system** - Various fire safety measures including automatic fire detection, automatic alarm system, automatic sprinkler system, PA system and automatic exhaust system were provided. Provision of water curtains and pressurization of lift shafts and staircase was also part of the design. Building was divided in five sections for fire detection purpose. Heat detection system provided in areas without air conditioning. Each hall/ room provided with two sets of detectors. Fire control room located at ground floor and manned 24 hours. Fire fighting system provided with (i) water tank and (ii) sprinkler system.

4.15 The building was divided into 32 smaller compartments for fire control purposes. This is done so that only a part of the building is disturbed in case of fire and not of entire building. Also, fire, smoke, gases and fumes should not travel from one place to another. Effective compartmentation done with the use of 'fire check' doors, with specified fire resistant fittings. The space between lintel and the ceiling, where clear space without services was available were sealed using 9 mm thick Calcium Silicate board on GI frame. The space where pipes, cables, cable trays were passing were sealed with two layers of mineral wool packing. Special attention was paid to ceiling of Electric Distribution Board (EDB) cut outs which were the communicating path from one floor to another. Fire resistant pillow of different sizes were used for irregular and smaller openings. The doors to AHU rooms were planned with 38 mm thick wooden panelled shutters. These were also made fire proof as well as sound proof by fixing plain asbestos sheet of 6 mm thickness on the inside. In addition, 40 mm thick mineral wool wrapped in 'markeen' cloth was filled between the asbestos sheet and newly made perforated aluminium sheet suitably fixed on wooden frame on the inside.

4.16 At one of the typical locations, an innovative solution had to be devised for meeting with fire safety requirements. The architectural concept planned that some natural light should reach the two basements to avoid sick building syndrome. Accordingly, glass block panels in combination with granite strips were provided all along the periphery of the building and periphery of courtyards to act as light wells. It was a novel concept of admitting natural light to the basement areas. However, intercommunicating space between the two basements disturbed the scheme for compartmentation. In this area, water curtains with the help of continuous sprinklers were provided as an effective fire fighting arrangement.

5. MISCELLANEOUS

5.1 Smoke extraction systems were provided in different manner in different areas. These areas were (a) car parking area; (b) Service Areas; (c) Area under focal dome; (d) Main building other than focal dome area.

5.2 Following type of lifts were installed; (a) Six number passenger lifts; (b) goods lifts – 4 number; (c) Book lifts- 10 No; (d) One capsule lift, with ten person capacity provided in library block; (e) Five person lift provided in auditorium for VVIPs.

5.3 Door frame metal detectors and X-ray baggage scanners were provided.

5.4 CCTV for Surveillance and library operation system provided along with CCTV (Display) system.

5.5 Acoustic analysis was done for Library and ten Committee rooms and acoustic treatment was provided accordingly. Sound reinforcement system was provided for auditorium and three committee rooms. Simultaneous interpretation system was provided in auditorium and three committee rooms. Film projection system provided in Auditorium. Main auditorium and seven Committee Rooms were provided with video projection system.

5.6 For announcement, car calling, quorum bell and music, public address system was provided. A comprehensive car parking management system was evolved.

5.7 Diaphragm wall was not to be punctured. Many services needed passing through diaphragm wall. So, puddle sleeves were provided in diaphragm wall at appropriate locations during construction. During excavation, earthing system was provided.

5.7 Exit light system and ceremonial lights were provided.

6. PROJECT MANAGEMENT

The project of this nature having numerous items of activity required a robust system of project management and a strong, and dedicated Project Manager. In the foundation stage, the work was broken down into a number of activities and targets were fixed. Actual progress was measured on fortnightly basis and corrective action taken as required. On account of such control, the work of diaphragm wall, retaining wall and huge excavation of about 2 lakhs cum could be completed in two years' time. For super-structure activities PERT chart was prepared, specifying monthly, weekly and for a few items even daily targets. From the year 2000 onwards, pace of work increased as number of activities multiplied many folds. Detailed programme of work, incorporating electrical, mechanical services and furniture was prepared using standard MS Project software. The number of active items to be controlled was more than one thousand three hundred. It was obvious that the entire review with concerned staff and contractors could not be completed in one sitting. Weekly review in parts was carried out covering about one fifth of items in one sitting. For the work of domes which was independent of other activities involving more than two hundred items, a separate MS Office programme was developed and monitored. Towards the end, a simpler but more detailed programme was prepared. List of pending activities, room wise, floor wise, block wise was prepared and monitored on weekly basis. The building work was substantially completed in January 2002.

7. QUALITY SYSTEMS CERTIFICATION UNDER ISO 9002

This being a prestigious and high-tech project of CPWD, a robust quality assurance system was necessary. Internationally acclaimed ISO 9002 Quality System provides a model for quality assurance and was considered essential for the project. A model for quality assurance in production and installation containing 18 elements of quality control requirements and methodology adopted to implement the same was required. Procedural requirements like formulation of basic documents were completed and submitted to the BIS who were accredited for such certification. Licence was granted for a period of three years from September 1996. This was further renewed for three years up to August 31st, 2002. The quality policy, expressing the overall intention of the top management endeavoured to ensure that quality objectives were met. This was the first project of CPWD constructed as per ISO 9002.

8. MAINTENANCE SYSTEM AND HOUSEKEEPING

A unique feature of this project was that a schedule of regular maintenance, along with description of the activity was evolved before completion. It specified, for example, 'French Sprit Polish and Melamine coating' on doors once in three years, 'Cleaning and disinfection of underground and overhead water tanks' once in three months, comprising dewatering, silt clearance, fungus removal, scrubbing, washing by pressure jet, chemical treatment and ultra violet radiation to kill bacteria and germs. Schedule of regular inspection for different components of the building like door, windows, fire check doors, floor cleaning, water supply fittings, cleaning of terrace drains and strainer pipes, to name a few, were specified so that in case of any deficiency, it could be rectified in time. Similar detailing was provided for furniture items and the kitchen equipment.

9. The work of housekeeping after completion was started in routine manner by manual cleaning. However, within a few months' time the Lok Sabha Secretariat decided that housekeeping work will be done by CPWD. It was also decided that mopping, cleaning activities should be done by machines, instead of manually. There were no established guidelines available then for automated cleaning. Therefore, a detailed study was carried out by CPWD in consultation with the equipment manufacturers, agencies carrying out such work and actual observations through demonstration. Equipment of standard make, having strong backing of after sales service were selected. Similarly, study was carried out about cleaning chemicals and detergents for each and every finishing material used in the building. Having decided the equipment, and the chemicals, next stage was to decide the frequency and the manpower required, including supervisor and manager, to carry out the house keeping work. A detailed list prepared for internal as well external cleaning and was followed.

10. Process of obtaining approval from bodies like NDMC, CE Elect NDMC, DUAC, Delhi Fire Service was initiated and obtained well in time. The building was completed in January 2002, and inaugurated by the Hon'ble President of India on 7th May, 2002.

11. Central PWD Engineers from Civil, Electrical and Horticulture wing worked hard for this prestigious project. Officers of Chief Engineer rank who worked tirelessly were S/Shri Jagmohan Lal, J N Bhavani Prasad, K Keshvan, Krishna Kumar, K Srinivasan, and K Ananthanarayanan. Civil Engineers include S/Shri Bipin Chand, A K Bajaj, R K Soni and Shri AK Garg Superintending Engineers; Sudhir Kumar, N K Garg, Jitendra Kumar, R K Duggal, R K Shami, and Sanjay Gupta Executive Engineers. Electrical Engineers include S/Shri S C Khurana, Mohan Swaroop and N Nagarajan Superintending Engineers; I J Malhotra, MML Chibba, G L Kapoor and SP Sehgal, Executive Engineers. Those were supported by dedicated and hardworking Assistant Engineers, Junior Engineers and office staff, for implementation of this project. Special mention is for Design Engineers, S/Shri Jose Kurien and Shailendra Sharma of CDO, CPWD.

LESSONS LEARNT

(i) **Customer Delight:** It was life time opportunity to build a monumental project like Parliament Library. Successful completion of the work to the satisfaction of the client was a matter of great pride and satisfaction. CPWD has brought out compilation of activities relating to planning, construction and maintenance of prestigious and monumental Parliament Library Building. This compilation is in two parts, one covering the Civil Engineering part of the work, and the other covering Electrical, Electronic & Mechanical Services in the Project. Books were published by the respective Chief Engineers, Civil and Electrical of the Project. Copies of the compilation were distributed to a select people. These compilations contain priceless information on innovative and uncommon construction techniques and required to be disseminated to Engineers and Architects engaged in planning, design and construction activities.

However, what is more commendable is that in these books, the documentation by Engineers is for each and every detail of the process of planning and execution of a project of such a magnitude Thus documentation provides a great learning opportunity to successive generations. It also throws up challenges and encourages to deal similar situations in a different way making use of contemporary technology. This practice requires to be followed for other major projects as well.

(ii) **Adoption of Hi-tech Construction Methods:** The compilation highlights the application of available technology and knowledge in resolving complex situations, like use of diaphragm wall with inclined rock anchors for excavation in confined situations, countering buoyancy due to high water table by using vertical rock anchors for the raft foundation, making AHU doors fire proof, making use of improvised machine for circular stone claddings etc.

(iii) **Committed Leadership and competent Technocrats:** All the structural design for the building, and design for electrical & mechanical services were done by the internal team of CPWD. Yet, there was no hesitation in obtaining expert advice from professionals outside dealing with it. The experts and the agencies engaged for specialized jobs were prequalified.

(iv) Proactive action on the following helped in ensuring defect free construction.

    (a) Preparation of mock up for the dome-cladding, and other finishes and sorting out deficiencies;

    (b) Laying out method statements, test requirements in great detail followed by thorough and timely checks. Defective material or construction was required to be removed forthwith;

    (c) Timely advice was given for speeding up the process, e.g., increasing number of moulds, steam curing, pre-casting of chajjas for domes etc.;

    (d) Judicious decisions were taken for identifying agencies and methods of waterproofing and making corrections in methods in case of expansion joints, roof treatment etc;

    (e) Frequent monitoring and sorting out co-ordination problems between agencies involved in the work could expedite execution of work.

(v) **Team Success:** Building confidence among team members and keeping their morale high was most important and it was done with great success.

Parliament Library Building -
Stone Railing & Interior View

Parliament Library Building -
Vertical glass block wall & Dome

Parliament Library Building -
View of Library

Parliament Library Building -
Rooftop view with the Parliament in background

1. BACKGROUND

During 1940, the Delhi Polytechnicwas established by the Government, at a historic building near Kashmiri Gate to cater requirements of industry. It offered Diploma courses in Engineering, Arts, Architecture, Textile, Commerce and Applied Sciences. From 1952 onwards, degree courses were started in this institute and was named as Delhi College of Engineering (DCE). Over a period of time the Department of Architecture become School of Planning and the Department of Art become College of Arts and moved out to new complexes. The Department of Chemical Technology and Textile Technology shifted out and made beginning of IIT Delhi. The Department of Commerce was abolished and faculty of Management studies of University of Delhi established. Under the aegis of this college, Netaji Subhash Institute of Technology was also established. During 1963, the Delhi Polytechnic was upgraded as Delhi College of Engineering and continued under University of Delhi at Kashmiri Gate. With passage of time the activities of college gradually increased. During 1980, the available space was much less than requirements. It was decided to get land for the Institution, to cater for the need of immediate requirements and for future. The land allotted was in North West Delhi near Bawana Road. This land was adjoining to the canal supplying drinking water to Delhi. It had good natural landscape.

2. APPROVAL OF PROJECT

Immediately after allotment of land, the survey plan of the area was prepared. The broad layout of the complex was drafted and preliminary plans of some buildings were prepared by PWD. Municipal authorities approved these plans and thereafter the matter was taken up with Urban Arts Commission. In Consultation with them, it was decided by PWD to design the complex with innovative designs. Accordingly, for designing M/s Ajay Chowdhry and Associates, a private Architectural firm was engaged. They had adequate experience of designing educational complexes. The scope of work was to develop the whole area including services as also buildings in the complex. Plans developed by Architects were accepted by PWD and approved by Urban Arts Commission, in consultation with college authorities. The scope of development of land included the earth filling work in the complex, construction of roads and parks.

tree plantation work. The scope of external and bulk services included, water supply reservoirs and external lines, sewer lines, sewage treatment plant, outlet for sewerage, electrical bulk services, electrical substations, distribution lines etc. The buildings in the complex to be constructed were, (i) Academic Block, (ii) Administrative Block, (iii) Humanities Block, (iv) Mechanical Engineering Block, (v) Library Block, (vi) Computer Center, (vii) Central Workshop, (viii) Hostels for boys and girls, (ix) Primary School, (x) Guest House, and (xi) Residential Buildings for faculty and support staff. During 1991, the Administrative Approval and Expenditure Sanction of the complex amounting to Rs. 101 crore was issued by Govt. of Delhi. The foundation of project was laid by Dr. Shankar Dayal Sharma Hon'ble Vice President of India.

3.   DEVELOPMENT OF LAND

The layout plan of the complex was developed to include works related to Roads and foot paths, Compound walls, Parks, tree plantation etc. Salient features are as follows:

  (i) **Earth filling in whole complex** - As the ground was low lying, it was decided to fill about one meter depth of earth in the whole area. The earth was supplied by Flood Control Department of Govt. of Delhi.

  (ii) **Roads** – Along the inner periphery of the compound, Ring Road was constructed with bituminous concrete. Various buildings were connected to this Ring Road by Cement Concrete Roads. The joints in concrete roads were provided with Kota Stone Strips to add to the landscape and to avoid cracks.

  (iii) For main gate, a special design was evolved in the shape of flying bird. Shri Sahib Singh Verma, Hon'ble Chief Minister gave guidance for evolving this design.

  (iv) Water bodies with glass mosaic tile flooring were provided at several places.

  (v) Extensive horticulture work was done and trees were planted, which could give extensive shade.

  (vi) Compound Wall was provided around the complex.

4.   DEVELOPMENT OF EXTERNAL SERVICES

Brief details are as follows:

  (i) **Water supply** – As no municipal water supply was available, six tube wells of 125 meter depth were provided at suitable locations. An underground tank of 3 lac liter capacity and an overhead tanks of 2 lac liter capacity on 20 m high staging, were provided. Water supply distribution mains were laid from overhead tank for supply of water to different building in the residential area. For non-residential buildings, direct pumping was done from underground tank, to water storage tanks constructed on terrace of these buildings. Wherever considered necessary, these tanks were interconnected. It was found that there was high fluoride content in the water and getting water from municipal sources was very costly. In consultation with Central Ground Water Board, a new location near Yamuna

Canal was identified. The tube well at this place gave good yield and quality of water was according to requirements.

(ii) **Drainage** – For drainage purpose, a RCC box drain (2.5m × 3m) was laid along the periphery of the complex. The network of external drains for different building was designed for connecting to this peripheral drain. In consultation with Municipal Authorities, the outfall for the peripheral drain was provided for connecting it to village Nallah in nearby area.

(iii) **Sewerage** – A municipal sewerline was available nearby. The sewerage system was designed for connecting to this sewerline and necessary permission of municipal authorities was taken. The assistance of Plumbing consultant was taken for designing sewerage system.

(iv) **Electrical supply** – Four indoor electrical substation were provided for getting HT supply from Delhi Electricity Supply Department, for distribution by LT cables. To avoid cutting of roads, RCC pipes were left beneath the roads to pass cables.

5. BUILDINGS

(i) In the complex, following buildings were constructed – (a) Academic block, (b) Administrative block, (c) Humanities block, (d) Mechanical Engineering Block, (e) Library block, (f) Computer centre, (g) Central workshop, (h) Hostels for boys & girls, (i) Primary school, (j) Guest house and, (k) Residential quarters for the faculty and staff.

(ii) **Foundation of buildings** – Sub soil investigation revealed that soil was primarily loose silt with low bearing capacity. The Structural consultant recommended providing 20 meter deep driven cast-in-situ RCC piles, to rest on hard strata. These were friction cum end bearing piles. Depending on structural requirements piles of 450, 550, and 600 mm diameter were provided. It was a huge task and three piling rigs worked round the clock for almost one year. Soil had large concentration of chlorides and sulphates. To overcome detrimental effect on durability of piles, following precautions were taken (a) Use of Portland pozzolana slag cement, (b) coating pile caps and grade beams with bitumen and, (c) instead of lean concrete below the beams, no fines concrete (one part of cement and eight part of coarse aggregate) was provided to break capillary action.

(iii) **Academic Block** – The Academic Block accommodated space for faculty, laboratories and lecture theater. The building was curvilinear in plan with two separate wings at 55 m distance and connected only at third floor, supported by two Y shaped columns standing free up to 11 m height. These columns supported a 55 m long, and 11 meter deep beam at third floor. On the beams there were columns to support the top beams. In this block there were diagonal chords of N truss. Besides these were Y-shaped columns. Some members were spanning three stories. Structural members in inclined shape posed a challenge for concreting work as pouring of concrete in inclined members was difficult and there was fear of segregation. Additional precautions were taken including (a) Accurate and proper geometry was ensured while doing form work because it was difficult to rectify infirmities and wrong geometry, (b) form work was made in such a manner that it could be removed in sections in desired sequence, without damaging the surface of concrete and disturbing other sections. Care was taken that no piece keyed into the concrete, (c) top shuttering was provided while casting diagonal chords of N truss and inclined members of Y shaped columns, (d) Concrete was poured in diagonal chords of N truss and inclined Y shaped columns through slit/opening provided in top of shuttering at about 75 cm spacing to avoid segregation, (e) It was ensured that concrete placed close to its final position to prevent segregation, (f) Vertical flow of concrete was restricted to 0.6 m, and (g) Concrete was discharged into forms directly to avoid segregation. With careful supervision, the work could be done as per requirements.

(iv) **Administrative Block** – It had a plinth area of 500 sqm with a majestic porch. For the roof 3.5 meter deep RCC N type truss of 21 meter span was provided at second floor level.

(v) **Library Block** – This building had plinth area of 5500 sqm. It was semi circular in plan covering 24m large span by folded plate structural system, having U shaped beams of 2m depth. Natural light was provided through the roof. Besides, a sloping curtain wall was provided in reading areas of all the floors. The structural design of the sloping curtain wall in curved shape was enabled with structural steel members encapsulated with aluminum sections.

(vi) **Computer Centre** – Computer centre having 2800 sqm area located next to the open air theatre, giving good landscape. The small porch was located at triangular corner. Raised floor was provided to accommodate complex cables.

(vii) **Hostels for boys and girls** – For Boys hostel five blocks were provided to accommodate 180 students in each hostel. Open balconies were provided. A Girls hostel to accommodate 72 students was provided.

6. COORDINATION WITH COLLEGE AUTHORITIES

Under the leadership of Prof. PB Sharma, Principal, College authorities took keen interest in implementation of the project. Name of Prof. D Golder is also worth mentioning. Several committees were formed to monitor progress of the project. All drawings were discussed in these committees. These drawings were modified whenever considered necessary. Thus, construction activity could proceed without any difficulty, with mutual cooperation and confidence between college authorities and Engineers of PWD.

The shifting of college was done in phases in four years period, adding fresh batch of students in each successive year. The first batch shifted in July 1996 and last batch shifted in July 1999. Thereafter entire college started functioning from the new complex. It is worth to mention that in order to stick to the program of shifting, PWD Engineer worked very hard.

7. ARCHITECTS

For this project M/S. Ajay Chowdhry and Associates were Architects. They did outstanding planning and introduced innovations. Their services are highly appreciated.

8. PWD ENGINEERS

PWD Engineers at various level displayed a missionary zeal to resolve a number of challenging issues and deadlocks to accelerate progress of the work. The project was controlled by a Superintending Engineer (Civil). Besides, there were electrical engineers and horticulture officers in charge of their part of work. Special mention is necessary for S/Shri PS Chaddha and BK Chugh, Superintending Engineers, who did outstanding work, along with other civil engineers including Shri Gajendra Kumar, Davsha Singh and Rajesh Bagga Executive Engineer as also Assistant Engineers and Junior Engineers. The team of Electrical Engineers worked under Shri Arun Puri Superintending Engineer (Elec.). The Horticulture work was controlled by Dy. Director (Horticulture). They all deserve appreciation. S/Shri DK Malhotra and IM Singh were Chief Engineers in charge and their contribution is highly appreciated. Shri KB Rajoria was Engineer-in-chief. The structural drawings were got vetted from Dr. Thiruvengadam of School of Planning and Architecture. He did excellent work.

LESSONS LEARNT

(i) **Breakthrough Improvement in Project Design:** Delhi Technology University at the new complex is one of the most prestigious complexes amongst technical Universities. To make it as a special institute, planning was done with new Architectural concepts. The observation of Delhi Urban Arts Commission, played an important role.

(ii) **Enhancing Customer Satisfaction:** Difficulties of foundations, techniques of construction, coordination of services etc, were sorted out by Engineers. The proper planning was done and project executed for today and tomorrow. Both college authorities and engineers did outstanding work. It is a success story for educationist and engineers.

(iii) Team Success: The coordination with clients and ensure their full satisfaction was very important. The college authorities decided to shift different classes in phased manner, as per their specific requirements. At that time buildings were under construction. Still space was provided in these buildings to accommodate the requirement of college authorities. Thus, satisfaction of client was achieved in turn their appreciation. College authorities highly appreciated the work done by PWD Engineers.

(iv) Success of the project was on account of close knit coordination between Consulting Architects, Structural Consultants, Client and Project Engineers. On account of this coordination, a project of multifold magnitude could be successfully completed according to strict time schedule.

(v) Proof checking of intricate designs gave added confidence to the team of Engineers.

Delhi Technological University Complex -
Foundation stone by Vice President of India

Delhi Technological University Complex -
Swarn Jayanti Udyan

Delhi Technological University Complex -
Inclined columns in Civil Engineering block

Delhi Technological University Complex -
Administrative Block

CENTRAL GOVT. OFFICE BUILDING AT LUCKNOW

The office of Accountant General for a State is situated in the respective State capital. The Accountant General functions under the Controller and Auditor General of India (CAG). The CAG in the supreme audit institution of India, established under article 148 of the Constitution of India. The Accountant General audits & reports about accounts of the concerned State Government. The growing work load of this constitutional authority required more man power, and in turn, more office space at Lucknow. The CAG office decided to construct a new office building at Lucknow for the office of Accountant General. During year 2008 it gave requisition to Central PWD, responsible for construction of Government of India Buildings, throughout the country. The request included category wise staff strength expected in this building along with space for other functional requirements like conference halls. A guest house for visiting officers and staff in transit was also to be provided. The CPWD office at Lucknow, in consultation with CAG office New Delhi, approached Government of Uttar Pradesh (UP) for allotment of about 3 Acres of land to accommodate office building and the Guest house. A piece of land was allotted in the Institutional Area of prestigious newly developing Gomti Nagar of Lucknow. In the vicinity several important buildings were located, which include New High Court Building, Indira Gandhi Pratisthan, Sashastra Seema Bal Complex, Ram Manohar Lohia Research Institute and several non-residential office complexes.

Physical survey work of the plot of land was carried out and survey plans sent by Lucknow office of CPWD to the Senior Architect stationed at New Delhi. On the basis of requirements given by CAG office, Preliminary Plans were drafted by Senior Architect, in consultation with engineers at Lucknow. These were duly approved by the client department. The general description of the plans drafted is as follows: (i) Plot area- About 3 acres (11,922 sqm); (ii) Plinth area of the building- (a) office building-15,257 sqm, (b) Guest house- 1,869 sqm, (c) two level basement- 8,246 sqm, sufficient for parking of 369 cars; (iii) The ground coverage was 26% against permissible 30%; (iv) Floor area ratio was 1.39% against permissible 1.5%; (v) Besides, normal office space, two number Conference halls of 13.0x7.0m and 19.5x7.0m size were also planned; (vi) Structurally the building was planned as a six storied framed RCC structure and double basement for parking; (vii) Raft foundation was considered for the building at initial planning stage.

On the basis of approved plans, a Preliminary Estimate (PE) was submitted. The Controller and Auditor General, Govt. of India accorded, Administrative Approval and Expenditure Sanction (AA &ES) amounting to Rs. 96 crores. The highlights of the project comprised: Dholpur stone cladding on external surfaces, UPVC /DGU windows for low heat gain, rain water harvesting, barrier free entry, sensors for conserving water, sensor operated LED light for energy efficiency, fully air-conditioned office space for optimum comfort for increased efficiency of the staff and insulated roof top to reduce heat for saving energy. Internal finishes as appropriate for the use of the areas was also approved by the client. Thereafter, CPWD worked out detailed specifications and structural design. Tender documents were prepared for Civil and Electrical works separately and work awarded to contracting agencies as per departmental procedures. It was decided to call tenders for other building services such as lifts, HVAC, firefighting, & fire alarms, DG-set and Pumps etc. when stages of execution arise as per provisions made in the PE. Civil works including water supply and sanitary installations & internal Electrical installations (IEI) was awarded at Rs.45.23 Crore to M/s Era Infra Engineering Pvt. Ltd, New Delhi. The work was scheduled to be completed in 16 months with date of start and date of completion as 20.10.2010 and 19.02.2012 respectively.

Foundation work- The structural designs were prepared by the Central Design Organization (CDO) of CPWD. The arrangement envisaged two level basements with raft foundation for basement parking. Sub soil investigations reported sandy soil at double basement level. The safe bearing capacity was reported as 11.31 T/m2 at this level. Further analysis revealed that there was a high potential for liquefaction of the soil during an earthquake on account of vibrations. Due to liquefaction, the structure could be damaged by sinking. Liquefaction is a function of relative density of soil and as also percentage of fines. The likely damages would be excessive if N value of soil was less than 1.5, as was the case. Therefore, it was considered necessary to take corrective measures. Accordingly, soil parameter was improved by using vibro compaction method so that it could sustain load of proposed building under severe conditions. For improving soil condition bore holes of 15mm diameter using pile dug bucket were made and then filled with grit of 3mm to 8mm size in layers, which were vibrated with vibration tool. This method enabled compaction of loose soil and improved drainage. The spacing was decided according to sub soil condition. This activity of improving the soil condition took about 5 months.

Building work- Considering the layout of the six storied building with two basements, it was decided to have RCC framed structure. For the main building two level basements were provided which were meant for car parking. The roof top was insulted for reduction in heat gain by the top floor which in turn helped in low energy consumption for air-conditioning of the building. Disposal of huge quantity of excavated earth from two layered basement was carried out efficiently. Close pursuance and coordination were also made to get the structural drawings from CDO timely. During implementation of the project, difficulties were experienced on account of slow progress of work by contracting agency. Issues were: i) Financial condition of the contractor was not sound in the last phase of the project; ii) There was insufficient deployment of labors at site of work by the contractor; iii) A few brands of items forming part of the contract were not available in the market; iv) Works of subsidiary works including services were not awarded timely; v) Shortage of funds on account of low budget provisions and; vi) Issues related to contractual items and architectural drawings.

CPWD Engineers handled the execution of the contract very carefully and in a professional manner. To resolve day today issues, regular meetings were conducted. Frequency of site inspection was increased, and notices issued to the contractor wherever required. This improved the situation. Finishing work took little more time than scheduled because a few finishing items such as Floriana marble and monopolized finishing items in the agreement were out of the market. Such items were suitably substituted with the consent of the client. Besides daily monitoring was done. Moreover, Extra and Substituted item statements amounting to Rs 3.0 crores were executed in such a manner that completed cost of contract after execution of these items did not increase. Officers who contributed for successful completion of work were Sarva Shri BB Gupta & Ramesh Chandra, Chief Engineers, Rajesh Banga, Superintending Engineer (Civil), BL Khandelwal Superintending Engineer (Electrical) and S R Awasthi, Executive Engineer (Civil).

At one stage the department considered to rescind the contract due to slow progress of work. It was a very difficult decision whether to rescind the contract or get the remaining works executed from the existing contractor by handling contract in more interactive way. After considering all the pros and cons, the contract was not rescinded. This also saved time in recall of tenders for the balance work. The Engineers had confidence on their professionalism instead of exercising powers to rescind the work they continued with the same contractor.

Bulk electrical services like HVAC, electrical substation, fire alarm system, firefighting system, lifts, DG sets, UPS etc. were independently taken up by Electrical wing by awarding work to different agencies under different agreements. Thus, a number of agencies were involved at the worksite, each complaining against other for creating hindrances. This was resulting in delay in completion of work. This position was analyzed critically and hindrance free site of work was ensured for each and every agency. Besides, their issues were resolved at work site. Civil and Electrical wing both worked in very cordial and coordinated manner to resolve issues raised by different agencies. The additional effort by Engineers for proper coordination among the contractors and making expedited payments, the building work was completed. The work would have delayed had the remaining work taken out of hands of main contractor and tenders called to get another contracting agency.

To reduce the cash flow problems faced by the Contractor during the last phase of execution, the department decided to make frequent payment to the contractor even if value of work done was small. In all 54 bills were paid, out of which many bills were of value of less than one crore. The Department also took initiative to make payment of the material directly to the supplying firms /venders in accordance with provision of contract. Mobilization advance was given to the agency as provided in the Contract. Extra, Substitute, Deviation item statements were sanctioned expeditiously. This helped in completion of the work as scheduled.

Under the saving of main work proposal for furniture work was got approved from the office of Controller and Auditor General. Furniture work was got executed through a separate agency. This facilitated immediate use of a functional and furnished building.

Arbitration - On completion of the works, the contractor was dissatisfied in respect of payment made to him and disputes arose between contractor and CPWD. The Contractor approached the Chief Engineer for the settlement of disputes through arbitration and filed claims of Rs 54Crore The claims of the contractor were related to price escalation, additional overhead expenses, abnormal hike in market rates of materials, extra expenses against WCT and electricity, claim against incentive for early completion, claim against loss of profit/damages, claim against penalty under Clause 2, claim against Extra Items Statements, Substituted Items Statements and Reduced Rate Statements and interest. The Contractor did not accept the Sole Arbitrator appointed by the Chief Engineer in terms of the Contract and approached the Court for appointment of outside arbitrator. While defending the case, department has also raised counter claim amounting to Rs. 2.1 Crore to recover liability of levy of compensation imposed by the department. The Arbitration case was defended in a very professional and meticulous way by the incumbent Executive Engineer (Civil), Shri Seraj Ahamad. The Sole Arbitrator, a retired Judge, dismissed the claim petition of the contractor and awarded NIL amount to him whereas counter claim of Rs 1.9 Crore was awarded in favor of the CPWD.

The credit for successful completion of work from the stage of sickness to the appreciation level go to the then Chief Engineer, Shri Ramesh Chandra and his team including all staff of CPWD involved in this project and also to the Executive Engineer who defended the arbitration case with dedication. The efforts by Engineers of department were highly appreciated by Shri S K Sharma, Comptroller and Auditor General, while inaugurating the building.

LESSONS LEARNT

(i) **Team Work:** Team work is the essential component of any successful endeavor. In this case, the main decision making from the client side was the CAG office, Delhi. The Senior Architect was also located at Delhi. However, day to day coordination was to be with the AG office Lucknow. It is to the credit of the field staff at Lucknow that necessary details and clarifications were obtained in time and it did not become a hindrance to the progress.

(ii) **Planning for Objectives:** Contract management is not to be simply taken as applying Contract clauses religiously. The prime objective is to complete the project. Thus, the Contract clauses have to be read and used in the proper context and considering all relevant factors, and the overall objective.

(iii) **Financial Management:** Solving cash flow problem of the Contractor in different ways is an example of out of the box thinking which led to the successful completion.

(iv) **Project Management:** Sanctioning of the Extra /Substituted items in terms of the Contract provisions well in time, and defending arbitration cases effectively are essential components of the project management. This was done in an effective manner.

(v) **Planning risk:** Time of completion should have some bearing on time consuming activities like ground improvement, basement with waterproofing, number of floors to be constructed etc and should not merely be based on the cost of the work. Non appreciation of the soil test report at initial stages did become a factor for extra time required in completion.

Central Govt. Office Building at Lucknow -
Front elevation

Central Govt. Office Building at Lucknow -
A view of the building

Central Govt. Office Building at Lucknow -
Another view of the building

Central PWD has been in-charge of construction and maintenance of Central Government buildings at Delhi and other places in the country. Funds for the maintenance were provided for Annual Repairs and Special Repairs. The expenditure for annual white washing, day to day repair and maintenance of services, special repair etc were covered by these funds. In fact, maintenance was value adding activity. It was important because, maintenance was necessary for conveniences of end users that is for an office or a residence or any other building.

At Delhi, number of field units that is Chief Engineers In-charge of Zones, Superintending Engineers and Executive Engineers in charge of Division offices were working for maintenance. Till 1975, while some Divisions headed by Executive Engineers were dedicated to maintenance, at division level. Superintending Engineers at Circle level and Chief Engineers at Zonal level were in-charge of both maintenance and construction. During that time Shri VR Vaish as Director General of CPWD reviewed the distribution of work and decided that maintenance for VVIP and other important areas be put in one zone under Shri OP Mittal Chief Engineer New Delhi Zone. For most of the other areas, the responsibility of maintenance both for offices and residences was given to Shri K Ramavarman Chief Engineer (Food). In turn for civil maintenance Circle VI was formed and Shri KB Rajoria was posted as Superintending Engineer (Civil). For corresponding electrical works, Electrical Circle VII was formed and Shri MK Mathur was posted as Superintending Engineer (Electrical). Under Circle VI, there were three units; D, G & M Divisions. S/Shri Jain, Gupta and Dewan were Executive Engineers. Shri K Ramavarman gave directions for improvements in maintenance. He wanted that maintenance should be economical, systematic and for benefit of occupants. He desired to sort-out difficulties in the maintenance work. The innovative work done by Circle VI, has been described for residential and office buildings separately.

During that period, the responsibility of maintenance of hospital complexes was given to another Chief Engineer. Shri AK Bhatnagar was Superintending engineer in charge of hospitals. He did many improvements in coordination with Shri KB Rajoria of which a few details are also described here.

I. MAINTENANCE OF RESIDENTIAL BUILDINGS

Enquiry Office Complex- Different Colonies under the charge of D, G and M divisions included Sarojini Nagar, Netaji Nagar, Laxmibai Nagar, Kidwai Nagar, Kasturba Nagar, Lodi Colony, Chanakyapuri, Timarpur etc. Enquiry Offices (or Service Centres) of CPWD were located within these colonies. For the purpose of maintenance of buildings, the Enquiry Office is the centre for all activities. Enquiry Office is generally located within the colony to be maintained. Complaints of residents are received at the respective Enquiry Office. Assistant Engineers and Junior Engineers have their offices at Enquiry office and deploy permanently employed workmen for attending to maintenance work. They are known as work charged employees. Materials for maintenance are also stored at Enquiry Office. Workmen with materials as required are sent for attending to complaints of residents and thereafter they report compliances at the Enquiry Office. Thus, Enquiry Offices were nerve centre for maintenance activity. Generally, both civil and electrical wings function from the same Enquiry Office.

All the Enquiry offices of different areas were visited by KB Rajoria along with Executive Engineers. During visit, it was noted that most of the Enquiry Office buildings were in bad shape, on account of poor maintenances and upkeep. Besides, in the compound of Enquiry Office, dismantled materials were lying. Most prominent of these materials were broken Overhead Water Storage Tanks, Flushing Cisterns and Wooden Shutters. Besides, for residents who visit Enquiry office, there was no proper reception with seating arrangement. Even front desks where Enquiry clerk received complaints were not proper. Therefore, the first important task was to bring Enquiry office complexes in proper shape, which was done in time bound manner. This preliminary initiative improved the image of Enquiry Offices and the residents were happy.

Dismantled materials in Enquiry Office Complex – (i) Dismantled flushing cisterns were examined. It was noted that most of the flushing cistern were not broken but rendered out of service on account of leakage, which was due to wear and tear at the place where bell was falling on bottom of cistern. These cisterns were shown to Shri SK Singhal, Executive Engineer Mechanical Workshop Division, who agreed to repair by putting some welding material on worn out surface and then rub to even out the surface. The proposal was accepted and nearly 90% of cisterns were made re-usable by Mechanical Workshop Division. That too at a nominal cost. Cisterns were painted and stored at Enquiry offices after reconditioning for re-use. These Cisterns were used and proved to be as good as new cisterns. These reconditioned cisterns were shown to Shri Ramavarman, Chief Engineer during his visit to an Enquiry Office. He highly appreciated the initiative and wondered as to why in the past, anybody else did not do so. There could have been no better appreciation for Engineers working for maintenance under his leadership. (ii) Dismantled wooden shutters were inspected in depth. These shutters were removed because some members were broken and new shutters were provided in place of these shutters. Close examination of these shutters indicated that 60 to 70 percent members of these shutters were in good condition and could be reused. Reusable wooden members of these shutters were removed by dismantling and stacked, Broken members which could not be reused were declared surplus and disposed off. New shutters were then made from reusable members and some members from new wood were added wherever required. Most of this work was done by work charged Staff permanently employed by CPWD at Enquiry offices. As per specific requirements, additional carpenters and beldars were employed on muster roll to give assistance. Thus, what could have been auctioned as unserviceable material at practically no price, was reused. Besides, opportunity for doing additional work was given to the Work Charged staff. They were otherwise attending to complaints of residents only and underutilized.

Management of Enquiry Offices for Residential Buildings - According to normal practice, the Enquiry Offices used to receive complaints from residents and these were recorded in a Complaint Register. Residents use to visit the Enquiry office and get the complaints recorded in this Register. Besides they were intimating by phone. At times, Assistant and Junior Engineers received complaints directly from residents. All complaints received were given complaint number and this number was intimated to resident for future reference. At Enquiry offices for bigger colonies like Sarojini Nagar and Lodhi Colony, separate complaint registers were maintained for each Junior Engineer as the colony was divided under different Sections for the purpose of attending to complaints. Besides, all JEs used to maintain their separate Store for building material to be used. The area of colony was divided under JEs who used to work independently. This was not considered proper. Besides, attendance of complaints also suffered for want of availability of material, because there was no fixed policy for procurement of materials. It was also noted that after attending to complaints the permanent Work Charged staff remained underutilized for want of adequate work. Thus, it was considered desirable to improve management of Enquiry Offices. It was decided that in bigger Enquiry Offices, instead of dividing the colony under different Sections, the complaints of whole colony would be recorded in one Register only. All stores were also put under the charge of one Junior Engineer. JEs were made in-charge of building work, plumbing work, carpentry work, etc. for the whole colony. Thus, JEs were able to see their job more in depth. The procurement of materials also became better as one JE was made in-charge of Store. This administrative change improved the working of Enquiry Offices.

Review of complaints - On review of Complaint Register, it was noted that a number of complaints remained unattended on account of non-availability of materials. According to delegation of power by the Department, each officer had a defined financial power during a year. Normal practice was that material purchase cases submitted for approval to higher offices and they took considerable time for approval. As per the prevailing policy for purchase of materials, cases for approval were submitted to EE and after he exhausted his financial power the case was submitted to Superintending Engineer. This approach was not considered proper as cases were submitted only after stock of materials exhausted. Instead, it was decided to plan purchases, for individual Divisions, by working out total demand of materials for different Enquiry Offices. Out of this list materials of very small amount were removed.

This list was reviewed and materials to be indented from Stores Division were taken out of the list. The list of remaining materials was reviewed and for materials available on Rate Contract, direct orders were placed. Remaining materials in the list were to be procured from financial powers of SE and CE and quotation called accordingly. Cases were submitted for approval by Superintending Engineer, with his powers and by Chief Engineer within his powers. Thus, processes of all purchases were done so that most of materials are timely available at Enquiry Offices. For remaining materials of small value and required immediately, financial power of Executive Engineer was used, to a large extent and these were termed as priority purchases. For unforeseen requirements, again purchases made from financial powers of Executive Engineers and termed as crises purchase. This procurement was done by on the spot quotation. With this approach the availability of store materials improved to a great extent. In turn the efficiency of work improved because complaints could be attended without delay. After complaints were attended by worker, it was considered desirable to get attended complaints checked by Supervisors and Engineers. This helped in improving the confidence of residents and created satisfaction amongst work charged staff.

Analysis of Complaints and follow-up action - Visit to colonies gave understanding of actual state of affairs. Besides, the critical review of Complaint Register also gave insight to the suitability of specifications, short-coming of planning, etc. A few examples are given here to explain the importance of such review. (i) Size of drainage pipe- In one Enquiry Office, for type two quarters there were many complaints regarding choking of kitchen trap and flow in drainage pipe. Some individuals registered complaints very frequently because pipes used to choke time and again. On inspection, it was found that 50 mm diameter size pipes were provided for kitchen outlets. Small size pipe was reason for this choking. So, pipes were changed to 100 mm diameter size. Thereafter, there were no complaints. Residents were happy with the replacement of pipes. (ii) Breakage of Glass- The consumption of glass required for replacement of window panes was very high. At some Enquiry Offices, as much as 7% of cost of materials consumed was for glass. Still complaints used to remain unattended. After inspection of area where glass consumption was high, it was found that glass was breaking in pockets where children's playground was around. At these locations glass was replaced by plywood. The glass consumption reduced to a great extent. It was noted at other places that size of glass panes provided in windows was large, so the cost of consumption of glass for replacement was high. In some type 2 and type 3 quarters, 2 to 3 sq ft size glass panes were used. It was not a proper planning by Architects. Change of window shutters was not possible. So, to avoid breakage of glass, instead reinforced glass was used for replacements. This reduced the possibility of breakages. (iii) Mumty Shutters- Mumty shutters used to break very frequently and at times complaint was registered at belated stage when extensive damage was already done. On inspection of these shutters, two main reasons were found for breakage. Many residents were not locking these shutters and with wind, shutters were becoming loose and breaking. Besides, one side of shutter was exposed to sun and rain. The paint was not lasting for long time and the shutter used to get exposed to severe weather conditions. In turn the shutter was breaking.

A special drive for repairing, replacing and painting of these shutters was done. For painting exterior finish paint helped to slow down deterioration. In such situations Engineers acted on proactive manner without waiting for complaints from all affected residents.

Job Standardization in Residential Buildings - At different Enquiry Offices, complaints received and special repair requests were reviewed with regard to maintenance work to be attended. It was noted that these complaints and requests could be classified under some Standard Jobs. These jobs were further grouped under two categories viz. Major Jobs and Minor Jobs. Jobs for rectification of defects/deficiencies due to uses or breakage were classified as Minor Jobs. A few examples being choking of floor traps, replacement of washers in taps etc. For attending these Minor Jobs, the cost component of labour was more and the cost component of material was very less. It was not possible to correlate these jobs with items provided in Standard Schedule of Rates. For Major Jobs like replacements etc. the scope of work could be worked out from items provided in Standard Schedule of Rates. For example, "Replacement of door shutters", was termed as Major Job. It included several items of work, i.e., removal of old shutters, transportation to Enquiry office, manufacturing of new shutter, transporting to place of work, fixing, applying primer and paint, providing fittings and to ensure that shutter functions properly. A standard unit in terms of a "Major Jobs" was considered better alternative compared to describing the work in terms of Schedule Items, which was required for cost accounting and not for appreciation of the work to be done. After review of work done through departmental labour and special repair works, standard list of Minor and Major jobs was prepared for residential buildings. Thus, all work to be done under maintenance except white washing and painting, was covered under Minor and Major Jobs. The concept helped in physical appreciation of work to be attended for maintenance.

Satisfaction of residents for maintenance - Order of satisfaction for maintenance work was ascertained from representatives of Resident's Associations. By and large they were happy with the attending of repair work, i.e., Minor Jobs through departmental labour. But they were not happy with attending of 'Major Jobs' which were generally done by contractors. Residents felt uneasy and uncomfortable with contractor's workers who were strangers. They were of the opinion that all Minor and Major jobs should be attended by departmental labour. Therefore, for satisfaction of Residents, for Major Jobs the contractor's role was kept limited to supply or procurement of shutters, tanks etc. and departmental labour was deployed for fixing. It was noted that, on average for attending to Minor Jobs, i.e., day to day complaints workmen were not fully utilized. They were remaining idle for about 25% time. Therefore, they were deployed to attend to fixing component of Major Jobs. If required casual labour was employed as per job requirement. This approach system was appreciated by departmental workers, who got opportunity to produce original work besides day to day maintenance. In turn, the total output of workers increased to a great extent. Thus, for same expenditure incurred the output was much higher and in turn more satisfaction to residents.

Performance Appraisal - It was considered necessary to judge the efficiency of work done by field units by introducing on index for Performance appraisal. Following details of each sub division were worked out and reviewed every month (a) Expenditure on maintenance during the month including emoluments of work charged staff but excluding expenditure incurred on white washing, painting etc. and (b) Assessed market value of work done including all Minor jobs and Major jobs attended. The difference between Assessed market value and expenditure incurred was termed as "Variance Factor". It was worked out for each and every Enquiry office, every month. The value of variation factor was giving fairly good idea about working of at Enquiry office. The plus value indicated efficient working and proper material management. The minus value indicated inefficient and uneconomical working. By comparing value of "Variance Factor", it was possible to compare efficiency of working of different Enquiry offices. Thus, Engineers were encouraged for achieving higher efficiency and more output.

II. MAINTENANCE OF OFFICE BUILDINGS

The maintenance work done by different divisions for a number of office buildings was inspected and position reviewed with administrative officers in charge of these offices as also Engineers. Conclusion was that maintenance of office buildings was generally not up to mark. It was noted that the responsibility for caretaking in these office buildings, was that of department occupying the portion of the office building, whereas responsibility for maintenance was with CPWD. There was some gap in understanding the scope of caretaking and maintenance work. It was concluded that important reasons which could be attributed to poor state of caretaking and maintenance were: (a) inadequate caretaking staff, (b) lack of supervision of caretaking staff, (c) delay in communicating the complaints by some offices to CPWD Enquiry office, (d) poor response for maintenance by Enquiry offices and (e) lack of civic sense amongst occupants etc. Besides, at some places, poor maintenance was on account of improper selection of finishing materials and improper planning of buildings. Moreover, wherever different floors of the building or portions of the same floor were occupied by different departments, the problem of maintenance was more compared to occupancy by a single department. East block – I, RK Puram was occupied by different departments including the office of Circle-VI CPWD (in charge of maintenance). It was decided to undertake caretaking of this block by CPWD and a separate Junior Engineer was posted to look after this responsibility. The expenditure of caretaking was to be shared by different offices and they nominated their representative to coordinate with the Junior Engineer of CPWD. He was directed to inspect all common area every day as also get information about requirements of maintenance and caretaking from different offices. On the basis of this information, Jobs required to be attended were recorded by Junior Engineer, in-charge of caretaking. The work to be done by Civil and Electrical units of Enquiry Offices was then communicated to them. The position was reviewed daily by JE caretaking and Civil /Electrical wings reminded for unattended to maintenance work. With this system of continues monitoring, the overall condition of buildings improved to a great extent. Improvements were recorded and were propagated by publishing on Wallpapers. It was on unique initiative which was appreciated by all concerned.

Another important requirement to improve maintenance was to review suitability of existing specifications of materials and planning of services. Important observations worth mentioning are (i) White mosaic flooring was provided in toilets which was difficult to maintain and (ii) The design of staircase nosing was not proper and there were daily breakages. The Specifications and designs were modified for better maintenance.

III. MAINTENANCE OF HOSPITAL COMPLEXES

The maintenance of hospital complexes was critically examined, when Shri AK Bhatnagar was Superintending Engineer in charge for three important hospital complexes. These were Safdarjung Hospital, RML Hospital and Sucheta Kriplani Medical Collage and Hospital. The maintenance of hospital complexes was responsibility of CPWD where-as the caretaking was that of hospital authorities. Hospitals being public buildings, visited by thousands of people every day. Maintenance and caretaking of hospital complexes was necessary to keep these complexes at acceptable standard of cleanliness, hygiene and safety. The most important requirement for proper upkeep of hospital complexes was to have mutually agreed division of responsibility and effective communication between hospital authorities and CPWD units, (civil, electrical and horticulture). By and large maintenance work of hospital complexes needed more attention compared to residential buildings, particularly on account of the fact that these were visited by patients who were unwell. A few very important maintenance issues are brought out as follows.

Seepage in water closet (WC) blocks of Safdarjung Hospital- Most of the WCs remained choked and on account of this choking, there was seepage spreading on walls and floors around the shaft. It was found that on account of poor design of waste pipe system, it becomes a chronic problem. There were three WCs in a row at one floor and there were four floors. On inspection of waste pipes, it was noted that horizontal pipe was provided to connect all the three WCs and this pipe was extended to shaft for connecting to vertical pipe common for all the four floors. Thus, there was no free flow of sewage and choking of any one WC was making all the three WCs unusable. Besides, sunken floor and wall were getting dampness. An alternative design was adopting by providing three separate vertical waste pipes stacks. These pipes were connecting one WC of each floor. The water proofing of sunken floor was redone and small drainage outlet provided in sunken floors. Besides, these stacks were provided on external wall and not inside the shaft. Thus, interconnecting horizontal pipes were avoided and vertical pipes were made approachable from outside. Choking, if any, could be attended on outer face without difficulty. This changed design solved the problem of choking. Thereafter toilets functioned in proper manner.

Change in design of door bend cover- The bolts of door-bend cover for waste pipes were of iron and used to get rusted with passage of time. For maintenance work whenever door bend cover was required to be opened these bolts would break. Thus, cover could not be replaced after attending to choking and bends used to remain open. So, bends became constant source of overflow. The Engineers of Hospital evolved a new design of door bend cover plate which was fixed by using a small bar, attached to the cover by threaded rod and fixed by a nut. This new design of cover made opening of door bend cover convenient. It was much more economical and convenient compared to replacement of door bend time and again.

LESSONS LEARNT

(i) **Value based Perception on Maintenance Work:** Maintenance of assets is not a popular activity world over as there are no inaugural stones to be unveiled or ribbons to be cut. Nevertheless, it is an important component of any project for user satisfaction as well as longevity of the assets. It makes economic sense as well.

Issues brought out here with regard to maintenance of residential colonies, office complexes, hospital complexes etc show that there are challenges for Engineers in attending to maintenance and achieving user satisfaction. Besides, out of the box thinking and management skills are required to motivate Workers and Engineers and improve productivity.

(ii) **Analysis of root cause and plan for corrective action:** It is also observed that by optimizing resources, total output can be increased. The critical examination and analysis of maintenance problems lead to improved design for the existing buildings. The feedback was also helpful for design of new buildings.

(iii) **End user/Customer Satisfaction:** Orderliness and economy for maintenance of buildings gives satisfaction of accomplishment. A few examples given in this article will be helpful to engineers to find innovative solutions to maintenance problems.

(iv) **Recording and circulation of lesson learnt:** it is considered that lessons learnt are circulated to other units so that they can take advantage for improvement of maintenance.

(v) **Requirement to modify specifications** and instructions for planning at the headquarter on the basis of experience gained.

PART

II

ROADS AND
RUNWAY SECTOR

SIDDHARTH RAJ MARG NEPAL

This road was earlier known as Sonauli Pokhara Road. During early sixties, it was initially taken up for execution by Regional Transport Authority (RTO) of Nepal. It could not progress and was left incomplete by RTO. Government of India came forward to construct and complete this road. An Agreement was signed between Government of India and His-Majesty's Government of Nepal, on 25.08.1964, for co-operation in promotion of development of communications in Nepal, by taking up construction of this road. Estimated cost of the project was Rs 9.11 crores, and it was to be financed as a part of the aid made available by India to Nepal. The projected date of completion was 31.12.1968. Some engineers of Government of Nepal were also to be deputed to the project, for training and getting experience in hill road construction. To resolve issues in implementation of project, a Coordination Board was to be set up, presided over by Minister of Communication, Government of Nepal. The Chief Engineer of the project was to be Secretary of the Board. The land for the project was to be provided free from encumbrances by Government of Nepal. Besides, supply of timber, sand, stone etc. was to be provided free of Royalty charges. Residual assets of RTO were also to be transferred to the project. Import of materials, machinery etc. was exempted from licensing requirements. Besides, project personal and others working for the project were allowed free movement. Contractors were allowed to import and repatriate Indian Currency. In a way all facilities for proper working were provided in the agreement, including responsibility for security of personal by Government of Nepal.

The work for construction of Sonauli Pokhara Road was entrusted to CPWD by Government of India. At that time, Shri NG Dewan, an officer of Indian Engineering Services, was heading CPWD. Shri CB Patel was posted as Chief Engineer of the project and he was temporarily stationed at Delhi. In order to ensure implementation of project, in efficient manner, Shri CB Patel and later on Shri Balbir Singh Saigal who took over from him, selected best lot of engineers and Shri NG Dewan, accordingly issued their posting orders to the project. They ensured that whoever was posted, was relieved and asked to proceed for working on the project and reach Butwal the base town along road alignment, located on foot hills, for further posting to specific place. At that time journey to the project site was very difficult. First it was a train journey from Delhi to Lucknow, then from Lucknow to Gorakhpur. After that the third train journey was from Gorakhpur to Nautanwa Bazar, near Nepal Border. From Nautanwa

Bazar, the travel was by Jeep via Sonauli in Nepal up to Butwal and it was two days foot march on hills along Trade route to Tansen, an important town along the alignment. Further upto Pokhra, it was seven days foot march. From Butwal onwards, the road was to be constructed in hilly region. Important places on alignment were Tansen, Ramdighat, Wallong, Putlikhet and Pokhara.

Office of Shri Balbir Singh Saigal, Chief Engineer, Shri VA Khaire Superintending Engineer and Shri RA Khemani Surveyor of Works were located at Butwal. Offices of Executive Engineers were established at different places along the alignment, to ensure expeditious execution of the project. Divisions 1 and 2 were at Butwal, Divisions 3 and 8 at Tansen, Division 4 at Pokhara, Division 5 at Putlikhet, Division 6 at Wallong and Division 7 at Ramdighat. The Chief Engineer had issued instruction that Executive Engineers should suitably locate their offices and office of their Engineers in their jurisdiction. Besides, they should make arrangements for their stay at the place of posting. Thereafter, all Engineers established their offices along road alignment. Temporary residences were constructed near to offices. Some engineers were provided tents for their stay, because it was not possible to make even temporary accommodation. All Engineers and staff were living under difficult conditions.

To encourage and give direction to field Engineers, Shri Saigal Chief Engineer along with Superintending Engineer and Surveyor of Works, started inspection on foot, along the road alignment. The team of Chief Engineer visited all the places, under very difficult conditions, and sorted out difficulties then and there. These inspections helped Chief Engineer to understand and appreciate circumstances under which different Engineers were working. Besides, there was open discussions with Engineers of different level. Thus, proper communication was established and in turn mutual confidence developed. Inspite of difficulties, the moral of all the Engineers was very high. In fact, Chief Engineer developed a team work spirit amongst all the Engineers. Shri Khaire, Superintending Engineer was an outstanding engineer with knowledge of hill road planning and construction. He drafted and finalized all the specifications and geometric design standards on the basis of specification and geometric design standards for hill roads of Ministry of Transport. Therefore, all other details were finalized by Shri Khaire with assistance of Shri R A Khemani. Even standard designs of culverts were finalized.

Shri OP Goel was Executive Engineer, SPR Division–3 at Tansen. He worked by involving all Assistant Engineers and Junior Engineers. He solved all difficulties without any delay. Shri Goel, used to make working comfortable and convenient. He used to encourage all engineers as also office staff and was a model to follow. There were four Sub-divisions in Division 3. First Sub–division was at Charchare. The second at Dumre, third at Burtung and fourth at Tansen. Near Charchare, there was a deep gorge and heavy rock cutting was required. Shri Amerjeet Singh was Assistant Engineer at Charchare. Under his leadership, the team of workmen and Engineers worked very hard. They did rock cutting, round the clock and were able to develop jeepable track in minimum possible time. His capability of consistent hard work, was an example for others to follow.During rainy season there was unexpected and unprecedented land slide near Charchare. It disturbed the road alignment and at places road was completely washed away. But where there is a will there is a way. Under the leadership of Shri OP Goel and under the command of Shri Amarjeet Singh, all Engineers, contactors and laborer worked day and night. Besides Shri KC Punj Assistant Engineer (electrical) provided support of machinery Within a few days, new alignment was established and road again became jeepable.

The second sub-division was at Dumre and Shri HP Khanduri was Assistant Engineer incharge. His team also worked very hard. Before on–set of rainy seasons, the work of approach road to a culvert could not be completed. In order to do this work during rains, the road work area was covered with tarpaulins and thus against all odds, this work was completed. Besides, for connectivity, two culverts were also constructed. Near Dumre, the road was to cross a river. In order to make the road jeepable, it was necessary to have a bridge. To construct this bridge, the whole team worked day and night and before rainy season, the work of bridge was completed. It was an impossible task. The work was got done with the participation of small contractors. Further, up to Pravas village the alignment of road was along the river bank. Beyond, the road alignment was rising on hills. This reach was in Burtung Sub–Division under jurisdiction of KB Rajoria Assistant Executive Engineer. On foot hills there was a mango garden. In case the ascending alignment along hill would have been taken in normal way, a number of trees would have been affected and the mango garden would have been destroyed. During that period, conservation of environment was not considered very important. Still, it was thought that, if cutting of trees could be avoided, it would be a good gesture. Ten alternative alignments were considered. Fortunately, one alternative was found where-in there was no necessity to cut even a single tree and that alignment was adopted. Local residents highly appreciated this initiative by engineers. Further, up to Bartung the jeepable track was made by ascending along the hill. In order to have gentle slope, one hairpin bend was introduced. The work for the link road from Bartung to Tansen was in Tansen Sub Division under the charge of Shri HS Luthra, Assistant Engineer.

On account of hard work by Shri OP Goel and his team, the stretch of road, within this jurisdiction was made jeepable. Upto that time, the road from Butwal to Charchare was not made jeepable by Division 2 and jeep could not move on this stretch of road. Therefore, one jeep was dismantled and brought by head load to Charchare. Thereafter it was reassembled and used in reach of S.P.R. Division III. It was a landmark.

The progress of works, in Bartung Sub-division was much ahead compared to other reaches of road. In other reaches, jeepable road was being cut, whereas near Bartung the work of full width excavation, geometric design, lines and levels, curve layout, transition curve, etc. was being undertaken. A sample reach was completed showing details of final work to be done in whole road. In sample reach even soling and metallic work was done. Besides, work of drain, parapet and a culvert was also done.

When the road from Butwal up to Tansen was made jeepable, the Chief Engineer informed the status to Government of India and Govt. of Nepal. During that period, Shri Shriman Narain, a famous Gandhian, was the Ambassador of India to Nepal. He came for inspection of road and saw completed reach near Tansen. He was highly impressed by the work done by CPWD. He desired that balance work should be completed as early as possible. After a few months of his visit, Shri Balbir Singh Saigal, the Chief Engineer, was honored with "Padma Shri", the National Award. It was an appreciation for all those who were working for that project. The work of excavation, soiling, metaling, black topping, drain and parapet as also culverts and bridges further continued. Division 3 under the leadership of Shri OP Goel was always ahead of others and all the works under this Division were completed within stipulated period and thereafter the Division was closed.

While working for this project it was an opportunity to learn many useful things which were very helpful later on, in discharging official duty. Most important was how to work in a systematic manner to achieve the laid down targets and develop a good team spirit as also bondage among colleagues. All this was possible due to the great leadership skills of Shri OP Goel who tirelessly guided and brought out best from the team. Engineers and otherswho worked in the team includeS/Shri Jagat Singh, SP Ganu, S Chakravarty, Ganguli, Badlu Singh, HS Chawla, MC Verma, PN Kapila, PR Sachdeva and KB Rai Khanna, Junior Engineers. Office staff include, Shri Swaroop, Head Clerk, Shri NM Gupta, Accountant, MR Dhir, RS Sharma, Maharaj Kishan, ML Zutsi and Naseeb Chand.

LESSONS LEARNT

(i) **Nepal Government support during project implementation:** Govt. of Nepal was to provide land for the project, assets of earlier implementation agency and royalty free construction materials such as timber, sand, stone etc. Besides, free import of construction materials and machinery which was exempted from licensing requirements. Free movement of personal for the project was allowed. Construction agencies were permitted to import and repatriate Indian Currency. All these facilities made implementation of the project very convenient. Thus, a proper speaking agreement, clearly bringing out role and field of responsibility by all concerned, ensured that the project could be implemented in effective manner.

(ii) **Leadership and Management Commitment:** Top level management has a very important role to play for success of the project. The deployment of competent engineers is always an important pre-requisite for the project. Proper selection of competent engineers and thereafter ensuring that they reach the project, was the responsibility of head office. On account of strict administrative control competent engineers and other staff members, joined the project timely. So, it was a job well begun. The project was implemented under the leadership of Shri Balbir Singh Saigal, Chief Engineer who was stationed at Butwal. He was available at the project site and this developed confidence amongst all engineer working for the project. Secondly, at starting stage itself, he made a visit to the alignment for first hand appraisal of the working conditions. He met all the engineers at their place of posting. After appreciating difficulties being faced by engineers, he gave personal assurance for his support in resolving the issues. In reciprocation, it was an opportunity for engineers to show their worth and work wholeheartedly for the project. This mutual confidence was responsible for the success of the project.

(iii) **Technical competence and contribution of middle level management for Project design and planning:** Any project can be implemented properly only if people along the line know what to do and how to do. The technical input in the form of specifications, schedule, methodology, designs and drawings was provided by Shri VA Khaire, Superintending Engineer, assisted by Shri RA Khemani, Surveyor of works. The outstanding work by their team was responsible to provide required inputs for what should be done and how it is to be done. Foresight or advance planning was another important factor.

(iv) **Advance planning:** Shri OP Goel, EE, laid down certain drills during rainy season so that the constructed road does not get washed away due to blocked drains. This includes use of tarpaulin covering road work to complete the work of culvert during the rains, and efforts for alternate alignment to save large number of fruit bearing trees. These are but a few such examples of the foresight coming useful in this project. Thus, advance planning is another quality essential in a good manager.

(v) **Clarity of task and proper coordination between middle and Junior level management**: Implementation of the project was most difficult part. Normally, according to CPWD procedure and systems, estimates are framed, tenders are invited by Press notice Thereafter works awarded to contractors. This conventional approach was not possible as contractors of reasonable level were not available. So, project was implemented through labours on daily wages or petty contractors. Shri OP Goel, took the lead for effective implementation, initiative to use resources as available and took practical approach. In fact, another person worth mentioning is Shri Amarjeet Singh, Assistant Engineer, having personal initiative and courage. Most difficult rock cutting was done under his command. Thus, from management concept, a role model, made implementation of project possible.

(vi) **A rare example of Human bondage:** A human bondage was established by Shri OP Goel, with all engineers and others who worked with him, under difficult conditions. Examples of establishing such bondage are very rare and shows that besides official relation, there is something more to share. Management principles, cannot explain such bondage.

Siddarth Raj Marg Nepal -
Shri Shriman Narayan, Ambassador of India in Nepal, visiting the project

Siddarth Raj Marg Nepal -
Shri Balbir Singh Saigal, Chief Engineer- introduction with Engineers during visit

Siddarth Raj Marg Nepal -
Shri Balbir Singh Saigal, Chief Engineer inspecting road work

Pasighat in Arunachal Pradesh is at foothills on the right bank of river Siang. The connectivity of interiors of Pasighat Sub division as also Along, the capital of Siang District, was necessary for convenience of travel and development of the region. Along the bank of river Siang, Pangin and Yinkiyong, important towns of Pasighat Subdivision were located. Besides, road from Pangin to Along already existed, which further connected to foothills. Moreover, from Pangin to Yinkiyong also road already existed. These roads were constructed by Border Roads Organization. Thus, proper road from Pangin was necessary for connecting to interiors as also foothills. The distance between these two places was 47 miles.

In 1971, KB Rajoria took over as Executive Engineer Pasighat Central Division. He was given an immediate task to make the Pasighat - Pangin road jeepable, by undertaking hill side excavation, wherever necessary and constructing temporary log bridges on rivers. Besides, the geometry of the road was to be improved and work of black topping, hill side drains, parapet and permanent culverts was also to be undertaken. Permanent bridges on major rivers were not in the scope of work.

The task to make the road jeepable was taken up expediously and engineers were motivated to work hard. Besides, Engineers were posted along the road alignment at different locations. Shri GS Likhari Assistant Engineer and Shri NN Basu Junior Engineer were posted at Rangin, a place at 15 miles distance along the road. Shri SN Brahamachari Assistant Engineer was posted at Pangin. To appreciate and understand difficulties, the road alignment was inspected by KB Rajoria on foot with Engineers and Supervisors. At some locations, the longitudinal slope was very steep and not suitable even for jeepable traffic. Besides, at some other locations, width was not adequate for safe driving. The work to be done was identified and work awarded to petty contractors on Work Order. In most of the reaches, there was no hill side drain. This work was also awarded along with work of temporary culvers.Moreover, location of small Nallas identified and work awarded to construct log bridges. There were very wide seasonal rivers at 27th mile and 29th mile. Engineers informed that multi span log bridges on these rivers did not stay long during rains, because of high velocity and heavy discharge. For these two rivers, extra hill cutting was done to move up stream. The purpose was to reduces height of log bridges. Besides, log bridges were made dipping in the middle, to reduce height. Besides, the curved shape made these bridges more stable and in turn stronger.

At 37th mile Yambung Nallah was to be crossed. It was 40 to 50 feet deep and having hard rock on both the sides. The width of Nallah was about 60 feet. It was not possible to construct a log bridge across this Nallah. For bridging this gap, about 80 to 90 feet long tree trunks were required. No such trees were available in the vicinity. However, suitable trees were located along the road alignment, a few kilometres away from Yambung. It was decided to cut these trees and transport to the bridge location. Felling of these trees was done by local residents without much of difficulty. In consultation with local residents, it was decided to use services of elephants for transportation. Accordingly, two elephants were engaged. The log was tied with ropes to one elephant and the Mahawat directed the elephant to pull the log along the road. Whenever the log was dragged for some distance, due to curve on the road alignment, part of the log used to move out on valley side. The second elephant was deployed just behind the log. Whenever log was going out on valley side, this elephant used to bring back the projected portion of log on the road. By carefully working, logs were taken to Yambung Nallah. Seven logs were collected at the location. Thereafter, again with the help of elephants these logs were pulled across the nallah, by tying ropes. After properly locating all the logs, the wooden decking was done, to make the bridge. This bridge was not endangered even during heavy rains and was to last for a number of years.

With sincere hard work by engineers and assistance from local workers and contractors, the road was made jeepable during that working season, i.e., in early April 1971. The local public highly appreciated the work of CPWD. During subsequent years efforts were made to keep the road jeepable as far as possible. There were frequent landslides during rainy season, which were removed manually on priority and drainage system for the road was kept proper. Of course, bridges at 27th mile and 29th miles did not last during rainy season. The road remained jeepable up to 27th mile from Pasighat side and from 29th mile onwards on Pangin side. However, during dry season, it remained jeepable, as these rivers could be crossed through the river bed.

GEOMETRY IMPROVEMENT AND PERMANENT WORKS

Estimate was framed and approved by Arunachal Administration for permanent works on Pangin Road. Scope of work included: (a) To make 12 feet wide road and improve geometry of Road to MOST standards; (b) Undertake work of soling, metalling and black topping; (c) Undertake work of hillside drain, parapet walls, Retaining walls and breast walls as also permanent culverts. Before starting the work, the position of construction materials and machinery for the project was reviewed, as brought out here- (a) Bitumen in drums was arranged from Indian Oil Corporation on Rate Contract and got transported by rail. Cement was also arranged on Rate Contract and transported by rail. (b) The workshop of Division was having ten Road Rollers of 8-10

Ton capacity and twenty number of one tonner truck. Stone crusher was procured. The diesel to run the equipment was arranged from Dibrugarh and transported by Boat and (c) Sand and shingle were obtained from river bed and hard rock was excavated from hillside for soling, metalling, black topping as also concrete work. Thus, availability of all materials and equipment for smooth execution of the project was ensured. Work was primarily executed departmentally, with its own fleet of trucks and road rollers. A Central Workshop was established with a stock of spares parts for vehicles at Pasighat. Experienced Mechanics were employed. Any spare part not available in the stock was to be procured from Dibrugarh. One-way journey by boat was taking 8 to 10 hours.

Design of RCC culverts was adopted from the hand book of Ministry of Road Transport for Class A loading. It was decided that geometric standards approved by Ministry of Roads Transport, should be followed. Assistant Engineers and Junior Engineers were neither aware of approved standards, nor they had any experience for field work. Fortunately, KB Rajoria has adequate experience of construction of hill roads. For guidance to Engineers, a detailed document was drafted by him giving provisions of standards and methodology to achieve these standards on the road. This document was titled "Planning of Hill Roads". Thereafter, practical training was given to Engineers so that the work on road could be done in a systematic manner and according to approved standards.

Stages of work- The work was implemented in following stages: (i) Milestone posts as per MOST standard, duly painted, were provided along the road at 100 meter distance. Location of each proposed culvert was also painted on hill side. A register of culverts was maintained. (ii) Field survey of existing road was carried out and plans including L-sections drawn for each mile. On survey plans, straights and curves were designed and marked according to geometric standards. (iii) The geometric design transferred to field by laying straights and curves. To suit the existing road condition some variations were done from drawings. Hill side excavation, as required to achieve geometric standards was done. Filling on road was avoided as for as possible as experience showed this was not stable. Primarily the hills side excavation consisted of earth, earth mixed with boulders, soft rock and hard rock. Classification of excavated material was done by Assistant Engineer and checked by Executive Engineer. Proper stacks of hard rock were laid on the road and measured. All excavated rock was accounted for in the material register. This hard rock was utilised for soling and metaling work later on. (iv) Straights and curves laid on the road. The length of straights and radius of curves, decided, to suit excavation already done. Total width of road decided according to geometric standards and by considering extra widening on curves as required. (v) L-Sections taken along centre line and plotted. The proposed formation levels marked on the drawing and cutting filling required on the road surface calculated. The proposed formation levels transferred to field. The formation levels on inner and outer edges also worked out and cutting filling required on the road surface marked. The super-elevation, both at start and end of the curve was given as per standards.

Extra widening on curves was also marked. Thereafter bed was prepared. (vi) Soling, metalling and black topping was done on road surface, V shaped stone drain provided on hill side, Parapet walls, Breast walls and Retaining walls of RCC / stone masonry provided as required. For culverts, parapet was provided for both hill and valley sides from safety consideration.

The systematic work for undertaking the road work was taken up both from Pasighat side and Pangin side. The work proceeded gradually during early 1975 Shri KB Rajoria was transferred and Shri DS Bhatia took over as Executive Engineer. Thereafter Shri AK Sarin took over. During his period project was completed. During 1987 Shri KB Rajoria was back to Arunachal Pradesh as Chief Engineer. He visited the road and was very happy to see the condition of the road and particularly geometry achieved.

LESSONS LEARNT

(i) **Setting of objectives and planning to achieve:** The concept of providing basic jeepable connectivity with temporary cross drainages is to be visualised in context of hitherto (early seventies) unconnected areas of Arunachal Pradesh some half a century back. During that period areas were still reachable only on foot. Some places, in hilly terrain, were about 20 days foot march away. Government officials were reaching such areas with the help of porters, halting on the way in Inspection Bungalows, carrying their own ration. As the population was scanty, traffic on these roads was also very low. Administration was keen to provide connectivity so as to take education, health and other welfare facilities to such remote areas. Needless to mention, initial stage development was very challenging.

(ii) **Team building and identifying innovative solutions:** The hillside excavation for this important road was already done partly. The first requirement was to establish connectivity. To have full assessment of the status, entire reach was inspected on foot along with concerned field engineers. This also helped in building rapport and to have their views on how to proceed. Discussions were held, innovative solutions were found and available resources were deployed. Learning from how logs were moved by deploying elephants was of immense help in shifting logs for the Yambung Nalla bridge. This speaks of courage and desire to achieve objective. Engineers at ground level deserve high appreciation.

(iii) **Adoption of standards for road geometrics:** Another aspect brought out in this article is about bringing the road to geometric standards. Adoption of geometric standards is essential to reduce driving fatigue as well as for the safety of road users. There has to be effort to improve conditions wherever possible.

(iv) **Knowledge documentation as a resource for future engineers:** Generally, in specifications and Standards its mentioned that what should be achieved. But these standards do not describe how it should be achieved. For this purpose, middle level engineers have to play important role. K B Rajoria gave practical instructions, which were followed by junior level engineers to achieve objective. The document "Planning of Hill Roads" played an important role. This is an important requirement for implementation of projects and senior level/middle level engineers, must develop practical guidelines for field work. Only after knowing what should be done, how it is to be done, the junior level engineers can achieve the objectives. Thus, job related training is essential. Resource planning is very necessary, as was done in advance for taking up permanent works for this road. This gave satisfactory results without any hassle.

Pasighat Pangin Road Arunachal Pradesh - Metalling work in progress

Pasighat Pangin Road Arunachal Pradesh -
Log being pushed over on Yambung Nallah

20

YINGKIYONG DAMRO ROAD, ARUNACHAL PRADESH

During 1972, the Arunachal Pradesh Government decided to construct a road from Yingkiyong to Damro in Pasighat Sub-Division. Yingkiyong is situated on left bank of river Siang, about 200 km upstream of Pasighat. A road was under construction up to Pangin, along right bank of river Siang. From Pangin onwards up to Yingkiyong a road was already constructed by Border Roads Organization. Damroh, the biggest village of Arunachal Pradesh was situated 60 km away from Yingkiyong towards South-East, across river Yamne. In these deep interiors, at that time, villagers were still living their traditional life, not much influenced by present day civilization. The Preliminary Estimate was submitted by Shri KB Rajoria, Executive Engineer, Pasighat Division, for construction of this road. The Arunachal Pradesh Administration, accorded approval for construction of Yingkiyong Damro Road, at an Estimated Cost of Rs. 60 lacs. The scope of work was, (i) Five meter wide road; (ii) Hill side drain; (iii) Cross drains-five number per Kilometer; (iv) Timber bridges on Nallahs; (v) Timber Bridge on river Yamne; (vi) 50 cm wide hill side drain; (vii) maximum longitudinal slope restricted to 5% and; (viii) Cross slope of road kept towards hill side.

To undertake construction of this road it was decided to create a Sub division at Yingkiyong under Pasighat Central Division, where KB Rajoria was Executive Engineer. Thereafter Shri AK Kaneri was posted as Assistant Engineer and Shri SP Acharya, Shri Ranjeet Kar, Shri NN Mitra as also Shri PK Chakraborty were posted as Junior Engineers. The administration had office of Circle Officer (CO) at Yingkiyong and office of Extra Assistant Commissioner (EAC) at Mariyang near Damro. They were under administrative control of Additional Deputy Commissioner Pasighat. CPWD Engineers requested the Additional Deputy Commissioner to give suitable instructions to local administrative officers for smooth execution of the project. Besides, he directed Shri Aple Parme, Political Assistant, to work with CPWD. Shri Parme hailed from Damro Village. He mentioned that in order to have smooth implementation of the Project, the cooperation of local population was necessary. He explained that in Arunachal Pradesh, the village administration was controlled by a set up like Panchayat and its Members were called "Gams." The senior most Gam was called as "Head Gam". They exercised administrative control in their respective Village, including law and order issues. Thereafter Engineers visited important Villages on road alignment along with Shri Parme. He explained about Government's decision to construct the road. The first important Village near the proposed alignment was Simon. Shri Anking Tasking was Head Gam at this village. Further near the location where the road was to cross the ridge, there was another important village and Shri Pangkong Meo was Head Gam. The road was to be constructed up to Damro Group of Villages where Shri Sonar Lego was Head Gam. During meetings, all Head Gams and Gams assured CPWD Engineers of full cooperation, so long as the village life was not disturbed and cultivation fields avoided from the alignment. They were offered to participate in road construction work by taking contracts and this gesture was appreciated by them. Some young boys came forward to take contracts for hill side excavation work.

The field survey work was immediately started and proposed alignment was marked on the Survey of India topographical maps. Thereafter preliminary site survey was done and the feasible alignment identified. Wherever the alignment was passing through village habitation or cultivation fields, it was shifted uphill or downhill. In some areas, trees were titling in one direction in a few reaches. After detailed investigations, it was found that the upper layer of soil was slipping gradually and that was the cause of tilting. Alternative alignment was found to avoid such slip prone areas. Besides, efforts were made to keep alignment near to ridge line, as road was likely to be more stable in such reaches. A suitable place was selected for bridge on Yamne River and road alignment was adjusted to connect at this location. The final alignment was inspected and approved. Permanent markings by the Executive Engineer were given uphill and downhill. Thereafter, tree cutting was done. At site, the trace cutting was done along the alignment. Cross sections were taken by conducting field survey work, quantities worked out and detailed estimates framed, for according Technical Sanction.

For implementation of the project in deep interiors, option for award of work by tendering was ruled out, as class I or class II contractors were not available. It was, decided that workable rates should be fixed according to local conditions and work offered to local tribal people and other petty contractors. Some contractors working in adjoining areas came forward and mobilized resources. It is worth to mention that at Pasighat several educated young man offered their services to work as petty contractors for participation in this project. They did commendable work for implementation of project by mobilizing local resources. This region of the country has very heavy rainfall. The annual rainfall was about 500 cm. The working season was limited form October to April. Therefore, the period for implementation was hardly seven months in a year.

Depending on resources available with contractors, reaches of 25 meter, 50 meter, 75 meter or 100 meter length were awarded from Yingkiyong towards Damro. After any contractor completed work on his reach, he was awarded work on further reach. If contractor did not complete work in time, his work was measured and he was paid. Balance work was given to another contractor. There was competition amongst contractors to maximize the output. All excavation was done manually and blasting of rocks was done as per requirements. It was done with Gelatin and detonators. For expedition implementation of project, the work was spread in whole reach of road. In case construction machinery would have been used, the work could have been taken only from one end and it would have taken more time to construct the road. For each reach of 25 meters, the percentage classification of soft/hard soil and soft/hard rock was done on the basis of actual observation by the Junior Engineers and checked by Assistant Engineer. The classification was hundred percent checked by KB Rajoria as Executive Engineer. It was a big job but considered necessary to have uniformity. The payment was done for quantities worked out as per classification. During rains, the maintenance was necessary. It was decided that wherever contractors did part work, it would be their responsibility to maintain the road. They were asked to cut drain on hill side even on partly excavated road and provide temporary cross drains. This initiative saved the road. In some reaches there were number of Jack fruit trees and during rains, Jack fruits were falling on road in thousands. So, clearance and removal of Jack fruits was a big task. In order to have proper maintenance, Junior Engineers inspected the road alignment every day to ensure that it remained clear of any obstructions. It was decided to construct a timber bridge on River Yamne. Suitable location was selected to ensure that on both ends hard rock strata was available. Timber logs were dragged to bridge site by elephants and the bridge was constructed, which stood during rainy season.

For implementation of project, the team led by Shri AS Kaneri Assistant Engineer, worked very hard. He had quality of leadership and inspired all Junior Engineers and contractors. All technical problems were sorted out by him by taking responsibility on his shoulders. He could keep excellent relation with local tribal population. They extended full cooperation. All Junior Engineers were very sincere and worked day and night. During February 1973, the jeepable road up to Damro was completed. It was an impossible task and unbelievable accomplishment. The local people were very happy. The local population of Damro decided to celebrate this event and Engineers were invited to Damro. Convoy of Enginees was cheered all along the route from Yingkiyong to Damro. When they reached Damro, all villagers gathered to welcome. After all construction of the road was a life time accomplishment for them. They had reason to celebrate and all of them were rejoicing and dancing.

The information about completion of road up to Damro was sent to Shri KK Raddy Superintending Engineer at Shillong and he informed Shri KAA Raja, Honorable Lt. Governor. He was very happy to know and decided to inaugurate the road without delay. On 06.04.1973 the helicopter of Lt. Governor landed at Mariyong helipad, near Damro. There was grand reception and thereafter they travelled along the road. There was reception everywhere by local population and they reached Yingkiyong by evening.

Honorable Lt. Governor conveyed his appreciation for completion of project in record time. He declared that those worked for the project would be rewarded and their services would be recognized by the Government. In view of recommendations of Deputy Commissioner in consultation with the Executive Engineer, Gold Medal was awarded to Shri AS Kaneri Assistant Engineer and Silver Medals to Shri Ranjeet Kar, Shri PK Chakraborty and Shri NM Mitra Junior Engineers. Gold Medals werealso awarded to Shri Aple Parme Political Assistant to Deputy Commissioner and Shri Sunar Lego Head Gam of Damro. Silver Medals were awarded to Shri Pangkong Meo and Shri Anking Tesking Head Gams. In the history of Arunachal Pradesh, so many medals were never awarded for any other project. After all it was very difficult task accomplished in a record time.

LESSONS LEARNT

(i) **Planning for Development:** For development activities to proceed in remote areas of Arunachal Pradesh, the most important requirement was road connectivity. Therefore, during early seventies a number of road projects were started. In remote areas it was not possible to follow normal procedure that is framing detailed estimate, calling tenders and award of work to the contractors. Therefore, for hill side excavation of roads, petty contractors were employed. A workable rate was decided and small reaches were given to these petty contractors. This was fastest approach to establish connectivity. Engineers were required to mobilize resources by involving local tribal leaders, village heads and educated youths. Besides, interested petty contractors from nearby areas were also mobilized. Proper living facilities were arranged for labour near the project site for efficient productivity. Law and order situation was maintained to normal and food supply arranged, with the help of local administrative officers. Thus, overall environment was favorably created, so that work on the project could proceed properly.

(ii) **Customer/End-user Delight**: This road project was implemented in deep interiors of Arunachal Pradesh. For implementation of the project, local resource, mobilization was done by Assistant Engineer and Junior Engineers in most efficient manner. Besides, the Political Assistant, a local leader employed in the office of Additional Deputy Commissioner, played proactive role. Engineers worked very hard and completed work for 60 km long road in record time. The achievement was unprecedented and output much more compared to other units of CPWD. Thus, even Hon'ble Lt. Governor was surprised. In fact, accomplishment was more than his expectations. For engineering professionals, it was unparallel achievement. The local population was very happy with completion of this road as it was beginning of availability of other services in the interior areas.

(iii) **Understanding the context for effective management during project execution:** Most important management lesson learnt was finding means of implementation according to local circumstances and environment. Human courage is supreme and project was implemented as there was will. The delegation, appreciation and encouragement by top management were equally important. The support of middle management helped in sorting out difficulties during executions of the project. It was one of the most satisfying task accomplished.

Yinkiyong Damroh Road Arunachal Pradesh -
After inauguration of project, in the picture
Left to right
(i) Shri KAA Raja Hon. Lt. Governor, (ii) Shri MM Lal, DY Commisioner
(iii) Shri KB Rajoria Executive Engineer &
(iv) Gagong Apong Local Leader, who later become Chief Minister

Yinkiyong Damroh Road Arunachal Pradesh -
Hill side excavation work in progress

Yinkiyong Damroh Road Arunachal Pradesh -
Cartage of raft rock by hand trolly

Yinkiyong Damroh Road Arunachal Pradesh -
The day on which jeep reached Damroh -
Villagers enjoying Ponung Dance to celebrate the occasion

21

ISBT FLYOVER & YUDHISTER SETU ACROSS RIVER YAMUNA, DELHI

1. INTER STATE BUS TERMINUS

Delhi is situated between Haryana towards North, West and South and Uttar Pradesh towards East. Total area of Delhi is about 1500 sq km. The city is divided by river Yamuna and 30% of area of Delhi is towards East of river Yamuna. Five National Highways join the city. National Highway NH-1 is connecting Karnal (Haryana) towards North, NH-10 towards West to Rohtak (Haryana), NH-8 towards South West to Jaipur (Rajasthan) via Haryana, NH-2 towards South East to Agra (Uttar Pradesh) via Haryana and NH-24 towards North East to Moradabad (Uttar Pradesh). After Independence of the country, there was phenomenal growth of population of Delhi. For road transport, radial and ring roads were developed. The Ring Road became the main road connecting different colonies of Delhi. During early seventies, the inter connectivity within city of Delhi was ascertained and Interstate Bus Terminus (ISBT) near Kashmere Gate was established. This bus terminus was situated on Ring Road, along bank of river Yamuna. For interstate connectivity, it was noted that through Ring Road, NH-1, NH-24, NH-8 and NH-2 were connected. The road cum Railway Bridge over river Yamuna near Old Delhi Railway Station and near ISBT provided connectivity to UP across the river. Besides, during 1957, the barrage cum-bridge over river Yamuna at Wazirabad was constructed, towards North of ISBT. These two bridges were inadequate for road traffic to East Delhi as also for heavy traffic near ISBT. It was considered necessary to provide a new bridge alongwith proper and adequate dispersal arrangement without traffic signal near the newly constructed ISBT. Accordingly, traffic and other studies/investigations were got carried out and preliminary plans were finalized. On the basis of this plan, a preliminary estimate was prepared and submitted to Ministry of Road Transport and Highways. During 1983 an estimate amounting to Rs. 37 crores was approved.

2. STUDIES AND INVESTIGATIONS

(i) **Traffic Studies** - The Central Road Research Institute (CRRI) carried out traffic studies and projected traffic volume of 1,16,000 Passenger Car Units (PCUs) per day by the year 2000 at this junction;

(ii) **Hydraulic Model Studies** - The Central Water and Power Research Station (CWPRS) Pune, carried out hydraulic model studies. They determined axis of the bridge, optimum water way without excessive afflux, span arrangement and river training work with repercussions on existing structure due to construction of the bridge. For these studies one-hundred-year flood criteria was considered. The design discharge adopted was 3.5 lac cubic meter per second. For well foundation 50% higher discharge was recommended;

(iii) **Geotechnical studies** – These studies were carried out by Geological Survey of India. In these studies, it was brought out that at the proposed bridge site, there was no likelihood of any rock upto the depth of 50 meters;

(iv) **Soil Investigation** – Preliminary soil investigation was done by Central Design Organization (CDO) of Central PWD. Twelve bore holes at proposed site of the bridge and fifteen bore holes at the western approach site were drilled. These investigation reports were to be used for preliminary design. On the basis of these reports, it was decided that for the bridge, well foundations and for the approach flyovers, pile foundations would be provided. Detailed soil investigation was carried out with one bore hole below location for each well for the bridge and at the foundation level location for the Western approach flyover for detailed design.

(v) **Aesthetic studies** – For working out aesthetic requirements including landscaping plan, an Aesthetic committee was constituted. This committee gave recommendations about size, ratio and proportion of various elements of the bridge and flyovers. Besides, a landscaping plan for the whole area within and around proposed structures was prepared.

3. PRELIMINARY DESIGN AND DRAWINGS OF BRIDGE AND FLYOVERS

(i) It was decided for bridge and flyover projects, the contracting agencies would be asked to submit detailed designs and specifications. It was however considered necessary to develop preliminary drawings and designs. The conventional dispersal arrangement using circular clover leaf was not found feasible due to proximity of the river Yamuna on one side of the Ring Road and existence of multi storied building of ISBT and Quedesia Garden and other structures along the Ring Road on the other side. Also, it was not possible to lower the Ring Road due to proximity to the river. The deck level of the main bridge was decided on the basis of highest flood level (HFL), freeboard and structural depth of girders. This prevented the bridge carriageway joining the Ring Road directly. Besides, there was no space available for a U turn back to Ring Road and ISBT on the Boulevard Road, which further on led to old Rohtak Road and was very close to the Tis Hazari courts. The issue of dispersal of traffic on Western end was raised in the Technical Committee meetings of DDA when the proposal for the new bridge was put up for approval. After several rounds of discussions, brain storming inhouse sessions, an acceptable concept was evolved. The Western approach was designed and planned in such a way that there would be no traffic crossing anywhere on Ring Road and Boulevard Road. Thus, twin projects, one of the bridge over river Yamuna and the other of dispersal arrangement on the Western end of the bridge were combined into one;

(ii) Two light curved flyovers on the river side which provided access to the main bridge were about 250 m long continuous girders over 6/7 intermediate supports. The box girder was designed as curved with length of cantilever of the box being generally kept uniform. Such type of girders was provided for the first time in India and required considerable innovations in analysis and detailing;

(iii) The bearings for vertical loads were selected as to permit individual rotation against all axes and also permit inplan movements. Combination of "Pot" bearings, elastomer vertical slab bearings with stainless steel sliding and metallic pin bearings for resisting horizontal forces was used over the piers. The vertical forces were transmitted to the pier through the pot bearings. On top of this Pot bearing a confined PTFE (polytetrafluoroethylene) layer was provided to permit movement of the superstructure along axis of the girder. Guide bearings were provided on every third or fourth pier. Vertical neoprene slab on which a stainless-steel surface permitted sliding was used for this purpose. The vertical elastomeric slab also permitted rotation of girders in plan. These guides were more essential at the end of the curved girders to ensure that the longitudinal movement at the expansion joint remain along the axis of the curved girders. For each curved continuous girder only on one pier a pin bearing was provided to resist huge horizontal force acting on the girder particularly under seismic condition;

(iv) The 250 m continuous girder in the light curved structure required specially designed expansion joints capable of movement between +115 mm to −80 mm. In fact, in addition to this anticipated longitudinal movement at places where curved girder meets the straight eight lane flyover, a lateral movement of plus/minus 25 mm was also expected. For the longitudinal movement in the carriageway, a sliding stainless-steel joint was designed to ensure smooth movement and durability;

(v) The curved flyovers were considered continuous over intermediate supports for the analysis. While designing it was assumed that each span will be cast on temporary bearings, prestressed to take the dead load and made continuous later. For horseshoe flyovers it was seen that individual spans, being simply supported at ends, were not stable while casting. The overturning moment due to eccentricity of the center of gravity (CG) of the span about its support in longitudinal direction causes instability. This necessitated casting of adjacent spans continuous over the intermediate support in one operation and prestressing them to cater for the dead load before removing the shuttering. The two span units thus cast were made continuous with other spans by casting the diaphragm and prestressing two continuity cables.

5. CALL OF TENDERS AND AWARD OF WORKS

(i) **The main Bridge** – (a) The work of main Bridge was awarded on lumpsum basis to M/s National Building Construction Corporation Ltd (NBCC). The project was implemented on the basis of drawings and designs prepared by NBCC and approved by PWD. The main bridge was of 12 spans of 46.23 m each, with 3.5 m deep simply supported twin box prestressed concrete girders. Total length of the bridge was 552.5 m. The deck slab was acting as the main compression flange. Each box girder supported 4 lane carriageway of 14.5 m, 1.5 m wide footpath, and central verge of total 1.8 m width. Four lanes rested on circular hollow piers of 2.75 m diameter on single circular well. For functional and aesthetic reasons, pier caps were made trapezoidal in shape. Wells were designed according to provisions of IRC:78-1983. Well diameter provided was 8 m, sunk to a depth of 33 m below the low water level (LWL). The left abutment well was 11.2 m diameter taken to 52 m below LWL. Both types of wells were provided with bottom and intermediate plugs. Intermediate space was filled with sand as per standard practice. East side abutment was cellular box type structure resting on both the abutment wells. Western side did not require any abutment as the bridge was in continuation of straight flyover crossing the Ring Road parallel to the river at this location. The bridge supported two water pipe lines of 90 cm diameter below the central verge. (b) There were financial difficulties and M/s NBCC was not able to complete the work. The portion of work not executed by NBCC was completed departmentally by PWD.

(ii) **Flyovers and Approach Road** – This work was awarded to M/s Uttam Singh Duggal. The structural design was done by M/s STUP India on behalf of contractors and approved by PWD. The Western approach was designed and planned in such a way that there would be no traffic crossing anywhere on Ring Road and Boulevard Road. This was achieved by providing five grade separators and two slip roads. The straight eight lane flyover, with 1.5 meter wide footpath on either side,was in continuation of the bridge. The span across ring road was 48.5 meter with a clearance of 5.5 meter above Ring Road. It descended near Tis Hazari Courts. A gentle gradient of 1 in 35 was provided to cater to all type of traffic. Eleven spans were on stilts. Total length of this flyover was 518 meters. Two light curved flyovers on river side of Ring Road provide approach to the bridge. These were 7.5 meter wide carriageways without footpaths, partly on stilts and partly on earth fill. Two horse shoe flyovers were provided for crossing the Ring Road, one on Metcalf House (North) side and other on Nigam Bodh Ghat (South) side. These also have7.5 meter wide carriageway, suitably widened at curves, and supported on square pillars. Two slip roads were provided at ground level. Total length of flyovers was about 3.0 km. The flyovers were supported on 168 bored piles and 807 driven piles. The earthen guide bund on the Eastern side was almost 1 km long with base width of 42.8 meter and top width of 6 m. Stone pitching over granular filter layer and launching aprons were provided as per design. The Eastern approach road was 2.9 km long and had dual carriageway, each 4 lanes with 3 meter wide footpath on either side. The Central verge was 1.8 meter wide. The approach road reduces to 6 lanes between the marginal bund and the G T road. The earth fill embankment was 6 meters to 12 meter high. Water bound Macadam (WBM) and Bituminous Macadam (BM)was provided as designed. Further on, traffic dispersal was through widened Road Number 66 bypassing the congested Shahdara township. However, the work of flyover was left incomplete by contractors. The PWD after giving suitable notices, completed remaining work at the risk and cost of contractors. This remaining work was executed by M/s Afcons. The works of Eastern approach road, slip roads and guide bunds were awarded to different contractors and they completed these works timely.

6. QUALITY CONTROL

Special attention was paid to the quality control measures. Suitable specifications and acceptance norms were prescribed in the tender documents. Tube wells were bored into the river bed for obtaining water for construction and curing. It was tested once a month for its suitability. If found unsuitable, Municipal water was used. No volume batching of concrete was permitted. Site labs were established for conducting day-to-day tests on cement, aggregates, water, cubes etc. Records, as prescribed, were maintained. Unsuitable materials were removed from site. Tests which were not possible at site were conducted at outside reputed laboratories. For example, HT steel was tested in IIT Mumbai and Chennai. Neoprene bearings were tested at Haryana Govt laboratory, Sonipat, bearings were tested and certified by DGS&D. To achieve desired concrete finish, shuttering was manufactured with new steel plates of 3mm or 6 mm thickness. Rubber packing was used in plywood shuttering joints to prevent leakage of slurry during concreting. Lines and levels were checked before concreting. These measures resulted in good finish which did not require any rendering.

7. DIFFICULTIES ENCOUNTERED

A number of problems were encountered during execution. These were resolved using technical and managerial skills of the team members.

(i) The site for construction of the bridge as well as approaches were full of encroachments like jhuggis, boat clubs, bathing ghats, ice factories, school and religious places. All these were removed and resettled amicably as per Government norms with the help of DDA. Compensation for the standing crop was paid for the guide bund and Eastern approach areas as decided by the revenue officials. Large number of services were passing through the Western approach area. These were coordinated with respective utility agencies for their diversion or shifting as per requirement.

(ii) Occasional problems during sinking of well arose. Where tilt and shift exceeded the permissible limits, remedial measures were adopted. In case of one well where tilt had exceeded permissible limits, it had to be sunk for additional 2 m to ensure that stresses remain within permissible limits. When the rate of sinking became very slow or the well got stuck, measures like additional kentledge, casting of additional lift, grabbing were resorted to. For the well below pier P7 on downstream side, a kentledge of 400 MT alongwith use of water jets around the well to reduce friction and continuous grabbing for 36 hours had to be resorted to take the well to the desired level.

(iii) A bottom plug is essential for proper functioning of the well as a load bearing structural member. Theoretical sump volume, calculated by assuming a theoretical profile and actual profile by taking soundings is compared until a desired profile is achieved. The concreting is then started. Sump volume calculated using soundings is then matched with the actual volume of concrete poured during plugging to call it satisfactory or unsatisfactory bottom plug. In case of one well there was a large variation between the theoretical and the actual quantity. The freshly laid bottom plug was immediately dismantled using heavy chisels, concrete re-laid after making the sump in a proper way. Help of professional divers was taken for plugging abutment wells at a depth of more than 55 m. A team of eight divers was engaged for nearly a month to ensure that the sump was clear of any muck before plugging.

(iv) For the main bridge the Contractor's design at tender stage envisaged use of precast prestressed girders and use of incremental launcher. During implementation, launcher could not be arranged for putting precast girders in place. A revised design with cast in situ box girders was then proposed by the Contractor and approved. There was delay on this account.

(v) Under the first span from Western side there was an all-weather flow of water with average depth of about four meters. It was designated as 'water span'. Several methods were considered for providing staging to support box girders during casting this span. The contractor, decided to drive 300 mm diameter temporary steel piles in the river bed to a depth of 18 m. Steel RSJs were then placed over piles. It took considerable time. Staging was then erected for casting of this downstream side span. It took more than one year to extract these temporary piles from the river bed. When the work was done departmentally, the upstream side water span was constructed by diverting the flow to next two spans and filling the first span with earth. This was economical as well as faster.

(vi) The curved flyovers were assumed continuous over intermediate supports for the analysis. While designing it was assumed that each span would be cast on temporary bearings, prestressed to take the dead load and made continuous later. For horse shoe flyovers it was seen that individual spans were not stable while casting, if simply supported at ends. The overturning moment due to eccentricity of the center of gravity (CG) of the span about its support in longitudinal direction was causing instability. This necessitated casting of adjacent spans continuous over the intermediate support in one operation and prestressing done to cater for the dead load before removing the shuttering. The two span units were made continuous with other spans by casting the diaphragm and prestressing two continuity cables.

(vii) During prestressing operation, it was found that the manually operated equipment was not giving satisfactory performance. These were then replaced with electrically operated equipment to obtain satisfactory results. Good quality of sheathing and proper taping at the joints was used to prevent leakage during grouting. Special care was also taken to ensure that grout mix remains of uniform consistency.

(viii) Partial prestressing of downstream water span was carried out after a flash flood warning was received. The girder was cast during the month of May supported on the staging resting on temporary piles. As per prestressing schedule, the prestressing of the box girder was taken up after the deck slab was cast. As the flood warning was received, initial prestressing was carried out, without casting of the deck slab, in a few strands, so that the box girder could take its own weight, should any contingency arise.

8. OFFICERS AND STAFF OF DELHI PWD

Officers and Staff of Delhi PWD have toiled and come with innovative solutions and adopted new technologies available at that time. One very important feature of this project was that, after discounting the risk and cost amounts, there was no cost overrun to the project. Shri L R Gupta, the Chief Engineer, guided the team comprising Sarva Shri A Chakrabarti, S P Banwait, Superintending Engineers, S S Mondal, N L Singh and A K Bajaj Executive Engineers, All deserve appreciation.

LESSONS LEARNT

(i) **Project Conceptualization:** Considering that planning for this bridge started and concepts finalized during early eighties, when the new technologies were not readily available locally, this bridge had many firsts to its credit. These have been brought out at relevant places in above document.

This project marked the beginning of assigning importance to traffic dispersal mechanism at both ends of the river crossing.

(ii) **Risk Management:** In an effort to obtain most innovative and economical solution there was an increasing tendency in government departments to invite lumpsum tenders with contractor's design. However, most of the contracting agencies depend upon external consultants to carry out the design. In this particular case it was seen that there was abnormal delay in submittal of detailed design for approval and execution. A solution to this issue is to insist on deployment of prequalified design consultants with proven track record who were expected to have the necessary manpower to do the needful.

In this case the entire mobilization advance, in terms of the Contract, was issued on award of work. It was seen that the entire amount was not spent for mobilization on this project. It would be appropriate to issue mobilization advance in stages, linking it to achieving certain preset milestone in the tender documents.

Progress of work suffered due to inadequate mobilization of shuttering material and key equipment. Therefore, it would be appropriate if the quantities are worked out with reference to the time of completion (and programme) and minimum quantity stipulated in the tender documents.

(iii) **Training and Competence:** Projects of this nature include many works and items which most of the staff is not familiar with in their day to day working. For satisfactory performance it is necessary that the deputed staff undergo necessary training.

(iv) **Focus on Workforce Welfare:** Most of the activities on such projects is round the clock. Also, many tasks, once started cannot be left halfway and have to be completed in continuation. The supervisory staff should be provided with required transportation and resting places along with meals to ensure sustained efficiency.

(v) **Set example for future Engineers:** Such projects which have conceived many novel ideas in design and construction should be seen as exemplary work and be a part of curriculum for post graduate students specializing in bridge engineering.

ISBT Flyover and Yudhishthir Setu Across River Yamuna, Delhi -
Main bridge across river Yamuna – view from below

ISBT Flyover and Yudhishthir Setu Across River Yamuna, Delhi -
Photo showing ground level slip road for U turn on Ring Road, elevated horse shoe
flyover for crossing Ring Road on Nigam Bodh Ghat side.

ISBT Flyover and Yudhishthir Setu Across River Yamuna, Delhi -
Eight lane flyover across Ring Road-View from below

During 1989, it was decided by Govt. of Delhi that for Civil Engineering works, particularly flyovers, an Engineering Wing should be started in Delhi Tourism Development Corporation and it should be renamed as Delhi Tourism and Transport Development Corporation (DTTDC). Accordingly, an Engineering Wing was created and Engineers were taken from Central PWD cadre. The funds for these flyovers were to be provided out of revenue of the Corporation from sale of liquor. The credit for this initiative goes to Shri VK Kapoor,Chief Secretary and Shri OP Goel, Chief Engineer PWD at that time. To undertake flyover projects, Chairman, UP State Bridge Construction Corporation (UPSBCC) was invited. For expeditious construction, prestressed, precast technology was mutually agreed. The efforts for establishing a casting yard and cost involved could be justified only if a number of flyovers were constructed using this methodology. The requirement of UPSBC was that the land for pre-casting of beams should be given without any charges and atleast four flyover projects be awarded. Their request was accepted and work of four flyover projects was awarded. These were at Loni Road in East Delhi, as also at Chirag Delhi, IIT Junction and Peeragarhi junction on Outer Ring Road. The work of Loni Road flyover was taken first of all and described herein.

Geometric Design- the task of traffic study and geometric design of flyover was awarded to M/s CRAPHTS. The traffic survey was done by them on Wazirabad road and Loni crossing. The traffic projections were also worked out. It was proposed that a four lane divided carriage-way flyover be constructed along Wazirabad Road. The traffic on Loni Road and turning traffic should be kept at ground level and a rotary be provided at the intersection which could be converted to signalized intersection when traffic increases. The proposal for traffic movement during construction stage was also worked out by the Consultants. After examining the proposal, the Engineers of corporation approved it.

Casting Yard- the work of planning and design was awarded to M/s Consultancy Engineering Services (CES) by UPSBC. The work of proof consultancy was given to M/s Mahesh Tandon and Associates. The land for the casting yard was allotted by Govt. of Delhi near Bhatti Mines, nearly 26 km away from project site. Length of precast girder that could be conveniently and safely transported was worked out considering the transport to be used and geometry of roads in the congested areas. It was decided that 18 m long pre-cast girder should be provided. The work of casting was started simultaneously with other activities at the project site. Two parallel casting beds were provided, each for casting of four girders simultaneously in one day on 'long line' method of casting. In this method, same prestressing strand was running through all the four girders in one line. After placing of reinforcements and stressing of strands, casting was carried out one by one. Steam curing of girders on the casting bed was started after initial setting and removing side shuttering. This way concrete was able to attain about 75 percent of 28 day strength in less than 24 hours. This was, according to design, sufficient to transfer of prestress, when strands were cut and four girders separated out. The girders were then moved from casting bed for wet curing. The same casting bed was thus ready for next casting in about 36 hours. Since this was a new initiative, it was decided to conduct load test on a prototype girder, both as independent girder and the continuous spans. It was got done through IIT Delhi. The general compatibility of the assumptions and behavior of girders in the load test gave confidence to consultants, contractors and DTTDC Engineers.

Sub-Soil Investigations and shifting of services: Initial Sub-soil investigations were done at four locations. By and large the soil strata were uniform. Top 4m to 5m depth was filled up soil. Below this layer there was 10-12 m deep layer of silty sand. On the basis of these soil properties, 1200 mm diameter bored cast in situ piles were decided. The services such as water supply lines, sewer lines, electrical cables and drains were shifted from the site of flyover project to adjoining places.

Geometry of flyover & structural arrangements: Wazirabad Road was raised at intersection and provided with elevated cast in situ structure, 46.8 m long central span, connected by precast beams on both the sides. Ten precast spans were provided, followed by retaining structure 86.50 meter long on each side. The total length of flyover was 565.20 meter. The central span at the crossing had vertical clearance of 5.7 meters (obligatory minimum 5.5m in Delhi then) and horizontal clearance of 45 m face to face of pier cap, for traffic on Loni Road across the flyover.

Foundations: For main span and viaduct approaches 1.2 m diameter cast in situ bored piles were provided as foundation. In general, piles were provided as pier columns (i.e., columns on single pile). For piers to take care of services, more than one piles with bridging pile cap were provided. This way shifting of services could be avoided. Open foundation with spread footings were provided for retaining wall and abutment.

Superstructure: In order to reduce number of expansion joints in superstructure, continuity for all the 10 approach spans in each end of the central span was established. Two circular columns on each pier for the substructure of each 2 lane carriageway were provided to improve the aesthetics of structure. For central span, post tensioned box girder was cast in situ on steal staging. Three stage construction for soffit, web and deck slab were done followed by single stage prestressing. The overall depth of box girder was 2.0 m. There were 30 working cables of which 4 were terminated at the soffit level. Provisions were also made for 4 cables for future prestressing. CCL system of prestressing was adopted using 12 T 13 cables 75 mm diameter bright metal sheathing was used. To ensure proper fabrication of cables, standard profilers were also used. Prestressing was done by J-20 jacks. Grouting of cables was done using cement slurry, having a water cement ratio of 0.4, mixed with 0.25% of super plasticizer to get adequate workability. M-45 grade concrete with 53 grade cement was used for box girders. Concrete was produced in batching plant and transported through transit mixers. Super plasticizer was used to achieve slump of 80-90 mm. A specially designed facia was provided for aesthetic matching of box girder with the approach spans.

General arrangement and bearings: There were two separate carriageways, each 7.5 m wide for carrying the up and down traffic. Each carriageway for approach span comprised of ten precast pre-tensioned girders. Only specific piers were fixed and all others were free piers. Two abutments and all piers except P-10 and P-11 had two lines of bearing. One line provided support to the inner end of approach span, while the other line provided support to one end of the central span. There were four Pot bearing on each pier/ abutment for the continuous deck. The central simply supported deck for each carriageway was supported on four elastomeric bearings, two on each pier P10 and P11. Bearing system provided for continuous spans was (a) Fixed pier (i) four Number Pot bearings for vertical load; (ii) one guide bearing for transverse horizontal force and; (iii) two set (2×2) vertical elastomeric bearings for horizontal longitudinal load. (b) Other piers (i) four number Pot bearings for vertical load and; (ii) one guide bearing for transverse horizontal load.

Expansion joints: Use of slab seal type elastomeric expansion joints at the end of 180 m long approach span were adopted to ensure a good riding durable surface. Slab seal type elastomeric expansion joints were provided to cater for movement of 180 meter continuous span of superstructure. Two types of expansion joints, D-80 and E-120, were used. First category was provided at the abutment, while second one was at the junction with the central span box girder. Installation of joints needed high precision fixing of steel inserts to the correct line level and cross slope of the deck slab. The portion of deck, where these inserts were to be provided were cast separately with close supervision. Presetting of insert plates as per prevalent temperatures conditions was done. After concrete attained adequate strength elastomeric slabs were inserted one by one from one end. Each slab was fixed in position with the help of fixing studs and nuts. These were then sealed with polysulphide to prevent loosening of nuts as well as ingress of water.

Miscellaneous works

(a) **Wearing course** - Wearing course of 25 mm Bitumen mastic layer with an under layer of 40 mm thick dense asphaltic concrete was provided. The mastic layer provided leak proof surface and permitted smooth riding. Bitumen coated aggregates were placed on the mastic layers for anti-skid measures.

(b) **Signage System** - Modern and properly designed signage system with combination of an engineering grade retroreflective material for road sign and high intensity grade over-head signages were installed.

(c) **Drainage from stilted portion and retaining walls** - The pipes from stilted portion were taken to the under-side of the girders. Drainage pipes were embedded in the cross-diaphragm while casting. These pipes were taken on the face of piers in the appropriate direction of the flow of traffic. Fittings were provided to ensure proper cleaning of pipes during maintenance.

(d) **Railing** - A properly designed steel railing was provided for the entire length of the flyover. For better durability epoxy paint was applied on the railing.

Street lighting : Lighting on the stilted portion and retaining wall portion was provided by using a 9 meter high street light poles with double brackets fitted with 400 Watt luminaries. The designed illumination level was 30 lux. In stilted portion specially designed galvanized steel brackets were provided in the deck slab. For retaining wall portion, poles were embedded in the ground. The street light poles on the slip road provided were 11 m high with single bracket, embedded in the ground. Total lighting circuit was distributed in 3 phases to ensure part illumination in case of failure of one phase. Electro-mechanical timer was provided for auto switching on and off.

Land-scaping and maintenance feasibility : At construction stage an appropriate land-scaping plan evolved with schemes for utilizing the space below the flyover. Due weightage was given to the accessibility of different components of the flyover for maintenance, inspection and replacement.

This project was undertaken by the team headed by Shri P B Vijay, Chief Engineer. Besides, he was assisted by Shri A Chakrabarty Superintending Engineer and S S Mondal Superintending Engineer. Their hard work and sincere efforts are reflected in this project.

LESSONS LEARNT

(i) **Out of box Thinking:** Adoption of precast pretensioned girders for construction of flyovers in Delhi was a pioneering effort on part of the team entrusted with speedy construction of flyovers in urban situations. This project took nearly 22 months. It was quite creditable.

(ii) **Risk Management:** All the pros and cons were considered and even a proto type was got tested by IIT Delhi beforehand for validation of assumptions in design. Factory made precast girders ensured good quality. Placing these precast girders stude by side excluded the requirement of providing staging for the approach spans, resulting in saving of time and avoiding traffic problems during construction.

(iii) **Time and Cost Saving:** There were many efforts to avoid shifting of utilities by providing 'bridging pile cap', to save time, temporary disruption in services and cost as well. Innovative technologies were used by adopting the then available latest materials in saving cost of construction, as well as providing better riding comfort.

Loni Road Flyover with Precast Prestressed Girders -
Side elevation showing sloping elevated structure of Loni Road Flyover

Loni Road Flyover with Precast Prestressed Girders -
Cast in situ Central box girder of 45m span

Loni Road Flyover with Precast Prestressed Girders -
Space below sloping spans enclosed for use as storage/ office

Loni Road Flyover with Precast Prestressed Girders - Elevated corridor

23 PUNJABI BAGH FLYOVER AT DELHI

1. STUDY OF RING ROAD

Rising population of Delhi along with rise in income levels constituted several challenges to the Urban passenger movement. Manyfold increase in number of vehicles resulted in very heavy traffic on roads of Delhi. Delay at junctions became a common phenomenon. Thus, the simple solution of segregating the cross traffic in time through signals become ineffective and other solution of segregation in space had to be provided. To find solutions for traffic problems, a comprehensive study on actual traffic volume count on all the arms of intersections on Ring Road was initiated in 1989 by Shri O P Goel, Chief Engineer, PWD Zone 1. Besides, a working group was also constituted by Delhi Govt. towards setting out scope of work, periodical reviews of the work carried out and convey its recommendations to the Government for implementation of traffic improvement projects. For estimating deficiencies at different sections of the Ring Road, field data was collected in respect of creating realistic profiles for speed and delay, traffic volume, pedestrian volume, land use, parking, noise levels, accident statistics. The data collected was then analyzed. Development schemes, as spelt out in Perspective Development Plan (PDP) 2001, were kept in view while assessing future projections. This study was carried out by Shri D Sanyal, Consultant, and head of NATPAC Delhi. The title of the study was aptly called 'Making Ring Road a Freeway', i.e., to say, it envisaged a signal free movement by suggesting phased improvements at all of its 47 crossings. In the first phase, the study proposed three level grade separators at 4 locations, namely crossings at Ashram, Safdarjung (AIIMS), Dhaula Kuan and Punjabi Bagh. Subsequently, traffic studies were also conducted on the busy stretches of Outer Ring Road. This marked the beginning of decisions for construction of flyovers based on scientific studies and the concept of corridor improvements.

2. LOCATION

The crossing of Ring Road and Rohtak Road (National Highway 10) was one of the busiest intersections of Delhi identified in the Ring Road Study conducted by Delhi PWD. It was decided to construct a three-level grade separator at this place and name it as Punjabi Bagh Flyover, being in proximity of Punjabi Bagh Club. Though the concept of a three-level grade separator was considered simple, the planning and designing was complex due to site constraints. To start with one task team was constituted comprising the group of engineers from planning, design and construction wings. At site office for Punjabi Bagh Flyover, the team met religiously every Saturday for brain storming sessions. It was decided by the team to raise the Ring Road and take Rohtak road below ground level with a rotary at the ground level for the turning traffic. Elevated Ring Road traffic continued over the existing ROB on the Delhi Rohtak Railway Line. Besides, one underpass was provided for traffic meant for the Transport Centre adjacent to the Ring Road. Next came planning for the pedestrian movement, which was substantial. Simple way was to provide four pedestrian subways, one on each arm of the crossing. In such a case the subways would have to be located nearly 300 m from the center of the junction as the central space was to be used for right turning traffic. Besides the subways on Rohtak Road could have come up only at the end of the underpass ramp portion. It would have involved considerable walking. Also, for any diagonal movement, it would have involved use of 2 such subways. This walking distance was unacceptable for pedestrian movement. After due deliberations, one practical solution was arrived at, which was to provide a pedestrian plaza below the central rotary, and above the underpass. It reduced walking distance to a minimum for a safe pedestrian movement. This satisfactory solution for the pedestrian movement increased the depth of the underpass to about 12 m. It threw up two challenges – construction of an earth retaining structure for such a depth and providing adequate space for the movement of the traffic at congested junction. The other issue which required solution was how to prevent base of the underpass and the retaining wall from behaving like a RCC boat under the influence of subsoil water table. For this purpose, a team comprising S/Shri A Chakrabarti, H K Srivastava and S S Mondal, Superintending Engineers was deputed to visit Calcutta Metro construction site to find out what could be learnt from them to control water table for deep excavation. Besides, contract conditions of Calcutta Metro were also to be studied to understand technical conditions, specifications, requirement of equipment involved in deep construction works in urban area. Findings related to above issues were discussed with the design and planning wings. The solution was to go in for the diaphragm wall to act as retaining wall in the deep excavation portion. It was open excavation and left ample space for the movement of traffic during the construction stage. It also avoided expensive shoring work. Problems associated with deep excavation in a busy intersection were also reduced. Solution against buoyancy was to increase the dead weight without uneconomically increasing the size of structural members. The solution arrived at was to use iron ore. Thereafter detailed planning of this flyover project was carried out. Thus, ultimately the scheme, consisted of three levels for vehicular traffic and one level for pedestrian movement.

3.   FLYOVER

Six lane flyover, 828 meters long, was planned along Ring Road, consisting of 516 meters of stilt portion and 312 meters of embankment. It was dual carriageway of 11 meters each. The longitudinal gradient was kept at 1 in 30 and vertical clearance above Ring Road was 5.7 meters (minimum required was 5.5 m). The obligatory central span was kept 50 meters, with two 35 meters spans on either sides and further 18 m spans. Expansion joints provided were "Strip Seal" and "Modular Strip Seal". POT-CUM-PTFE Bearings were provided. The substructure was of RCC pier caps supported on RCC piers, which were supported on Bored cast-in-situ RCC piles. Concrete used was M-25 for piles and pile caps, M-35 for piers and pier-caps and M-40 for super structure. For embankment, Reinforced earth construction technology was used. Fly ash was used as fill material with Geo-grid reinforcement. M-35 grade concrete blocks were used for Reinforced Earth Wall. The crust thickness of road consisted of granular sub base, granular base, wet mix macadam, bituminous macadam and dense asphaltic concrete. Wearing Course of 25mm bitumen mastic was provided over 50mm Dense Asphaltic Concrete (DAC) in stilt portion and Micro Surfacing with polymer modified bitumen was provided on embankment portion.

4. MAIN UNDERPASS

(i) Six lane main underpass along Rohtak Road was constructed at 11.8 meter below the road level. The extra depth was for accommodating the pedestrian plaza below the ground level rotary. For such a deep structure, diaphragm wall technology was considered suitable. Salient features of main underpass being: (a) Total length – 626 m, (b) Covered portion – 100 m, (c) Trellis portion – 100 m on either sides of covered portion, totaling to 200 meter, (d) Open portion with Diaphragm Wall – 75 meter on either side totaling to 150 meters, (e) Open portion with retaining wall- 88 meter on either side totaling to 176m, (f) Vertical clearance – 5.5 meter, (g) Width – Dual Carriageway of 11 meters each, 0.7 meter Safety Kerb on either side and Central Verge of 1.2 meter, (h) Longitudinal Gradient 1 in 30, (i) Camber 2.5%. (ii) Foundation System for the underpass -For the underpass, an underground structure, the uplift force due to buoyancy was a major factor in deciding the foundation system. The depth of water table was varying from 6 to 7 meters. For design purposes, worst conditions were assumed by considering water table at 3m below the ground level. Besides, as a precautionary measure, weep-holes were provided. The foundation system consisted of RCC base slab spanning between outer diaphragm walls, with an additional intermediate support in different zones by way of diaphragm wall and piles respectively. The thickness of base slab varied from 500 mm to 800 mm. To counter the uplift forces on base slab due to buoyancy, iron ore mix was laid on base slab. It was a reliable and economical solution. The thickness of iron ore layer varied from 500 mm to 1300 mm. During construction, water table was kept lowered by continuous pumping. At finishing level of underpass, Rigid pavement 230mm thick with M-40 grade concrete was provided. (iii) Drainage System for the underpass -It was necessary to provide proper drainage system so that it does not get submerged and remain functional during rainy season also. For effectiveness, sumps were provided just adjacent to underpass. The sump walls were made by diaphragm wall construction of adequate depth and for required capacity. Two sumps were made on either side. The storm water to collect in sumps through storm water drains along the carriageway. Centrifugal pumps were fitted with water level sensors were provided in sumps. For design of sumps average rain fall was taken 25mm per hour, almost double of past data of rainfall; (iv) When the project was started, to support diaphragm wall construction, temporary steel struts were provided between two diaphragm walls. It was found time consuming job. So, design of diaphragm wall was modified by extra anchorage to avoid use of struts. This change facilitated expeditious execution of work.

5.   PEDESTRIAN PLAZA

Pedestrian plaza was unique concept provided for the first time in the country. The pedestrian plaza was covering entire intersection on Ring Road and Rohtak Road with arms of entry/exit on all the four corners. It also provided space for putting thirty shops, the area of pedestrian plaza was 2160 sqm. For proper identification by pedestrians, differently colored finishes were provided for each of the four arms of the pedestrian subway. Besides, forced ventilation system was provided. Two Fresh Air handling units were provided for supply of fresh air to pedestrian plaza. The total hourly air change was considered more than twelve. Acoustic lining and fire dampers were also provided. Supply of fresh air was planned through linear grills and outflow through arms of pedestrian plaza.

6.   ROTARY

Rotary at ground level was provided for turning traffic. Size of rotary was about 75 × 50 m. To serve local traffic, a box culvert as link road underpass was provided, north of main flyover.

7.   MICRO-SURFACING

The embankment portion of flyover was treated with micro-surfacing. The micro-surfacing was done on Dense Asphaltic Concrete surface. This micro-surfacing was a mixture of polymer, modified asphalt emulsion, mineral aggregate, mineral filler, water and other additives, properly proportioned, mixed and spread as a paved surface. It was done by a truck mounted unit, equipped with a continuous flow mixing unit. It had suitable means of accurately metering each individual material and was capable of delivering a predetermined proportion of aggregates, water, additive and asphalt emulsion as a mixture. This mixture was spread on the road surface with the help of spreading equipment attached with mixing machine. This treatment enhanced life of wearing course. Besides, its textured surface was anti skid.

8.   SHIFTING OF UTILITIES

Services belonging to different organization shifted during execution of the project were: (i)Three water mains of sizes 600mm diameter, 900mm diameter and 1100mm diameter, belonging to Delhi Jal Board; (ii) 33 KVA lines (overhead and underground), 11KVA lines and other cables belonging to Delhi Vidyut Board were shifted. Shifting was planned in such a manner that diaphragm wall work below the HT line could be executed; (iii) Telephone cables;(iv)Besides a number of trees on the alignment were cut/transplanted after taking approval of Forest Department; (v)Traffic signals were shifted, as per specific requirements of the traffic diversion plans; (vi) DTC Bus Stops were also shifted as required.

9. REQUIREMENT OF LAND ON TEMPORARY BASIS

No space was available near the site for contractor's office and Batching plant. Therefore, 3,000 sqm of MCD land and DDA land were taken on lease. For labour camp and casting yard, space was provided at about 12 km from site. Transfer of land took time and caused delay.

10. PRODUCTION AND TRANSPORTATION OF CONCRETE

Two fully automatic computerized batching plants of 30 cum per hour capacity were installed by contractor, at project site. Another batching plant of 15 cum per hour capacity was installed at casting yard for casting of I-girders. Six transit mixers of 4 cum capacity each were deployed for construction. A concrete pump was used for lifting concrete. A fully equipped laboratory was provided for batching plant. For this project about 30,000 MT of cement and about 7,600 MT of reinforcement steel was used.

11. LIGHTING

Fourteen high masts were provided, in conjunction with the street light poles for effective illumination. High Masts were 30 meter high and fitted with 8 number HPSV fittings of 2X400W. The illumination level was in the range of 30-Lux. The street lighting system was connected to PWD Electric Substation. Besides, standby Diesel generating sets were provided. High mast lights were made operational in the beginning of the project, so that round-the-clock construction could be done. Besides, it was useful for night traffic movement.

12. ISO 9002CERTIFICATION

 In order to ensure that project is implemented with highest quality standards, ISO 9002 certification was taken by PWD Delhi for this project. PWD Engineers took necessary training for certification.

13. CONSULTANTS, CONTRACTORS AND PROJECT PERSONAL

Initial conceptualization of the project was done by a team of Engineers including S/Shri H K Srivastava and S R Pandey, Superintending Engineers, as also S/S P C Srivastava and S C Arora Executive Engineers. The Traffic Engineering study and geometric design was done by M/s CRAPHTS. Thereafter, the detailed structural planning and design were done by M/s Consulting Engineering Services. Dr. SK Sharma Superintending Engineer was in-charge of the project and structural design were checked by him and Shri Jose Kurien Superintending Engineer, Central Design Organization, Central PWD. The work was awarded to M/s. Afcons, a reputed contracting firm and they did excellent work. During execution of the project, Shri KB Rajoria was Engineer-in-Chief PWD Delhi, Chief Engineers Incharge of the project were S/S KN Agrawal and Shri BK Chugh. Shri SK Rustagi was Superintending Engineer. Executive Engineers who worked for this project during execution were S/S HP Meena, Shri Pradeep Garg, Shri Mathura Prasad, Shri PC Jain and Shri Taneja. Assistant Executive Engineers wereS/Shri PS Rao and Shri Himanshu Pandey. Assistant Engineer were Shri AK Gupta and Shri HR Rohilla. Junior Engineer were S/S Soumen Bardan, Shri SC Jain, Shri Venkat Rao and Shri Satya Prakash.

LESSONS LEARNT

(i) **Challenges in Project Conceptualization and Design:** Evolving a user-friendly design within the available right-of-way and then providing for geometric and structural requirements was a big challenge for this project. Brain storming sessions at the project site led to the development of a four level grade separator, for the first time in the country. Team spirit achieved the objective.

(ii) **Customer/End-user focus:** Convenience of the pedestrian was considered of equal importance compared to motorists by providing for a pedestrian plaza all by itself for pedestrians which was well ventilated, lighted and each arm providing with different colors for ease of navigation. This also reduced walking distance. It was for the first time in the country.

(iii) **Use of Advanced Technology:** To reduce problem of motorists during construction due to space constraint, advanced technologies like use of diaphragm wall and geo-grid reinforced soil wall were adopted.

(iv) **Learning Best Practices of the Industry:** There was no hesitation in trying to learn from other projects in the country having similar work conditions. Dedication and hard work during planning and implementation stage led to successful completion of the project.

(v) **ISO standard certification:** For assuring construction with quality standards ISO system certification was obtained for this project. Field staff were suitably trained to ensure proper control.

Punjabi Bagh Flyover at Delhi -
Elevated Ring Road with Rotary at ground level

Ch-23 Punjabi Bagh Flyover at Delhi -
Six lane main underpass carrying uninterrupted Rohtak Road traffic,
Elevated Ring Road is seen in the background at right angles

Punjabi Bagh Flyover at Delhi -
Pedestrian subway below Central rotary

Punjabi Bagh Flyover at Delhi -
Both Carriageways in operation

AIIMS AND DHAULA KUAN FLYOVER PROJECTS AT DELHI

All India Institute of Medical Science (AIIMS) and Dhaula Kuan Crossings were busiest intersections on Ring Road of Delhi and movement of traffic was abnormally delayed at these junctions. At Dhaula Kuan, National Highway Eight (NH-8) joins the Ring Road and at AIIMS, Aurobindo Marg crosses the Ring Road. The traffic engineering study conducted on Ring Road, recommended three level solution for these intersections. The PWD prepared proposals for these intersections and got approved from Delhi Development Authority. According to procedural requirements, these proposals were submitted to Delhi Urban Arts Commission (DUAC), a statutory body, working under Ministry of Urban Development, Govt. of India. The DUAC examined these proposals and did not approve. Even modifications did not find favor with DUAC. Thus, there was complete stalemate and revising proposals would not help in getting approval.

During that period, Shri Madan Lal Khurana was Hon'ble Chief Minister of Govt. of National Capital Territory (NCT) of Delhi. The matter was brought to his kind notice by Shri KB Rajoria, Chief Engineer PWD. It was decided that a meeting should be arranged in his office where-in Chairman DUAC should be requested to join to sort out the difficult situation. The meeting was according organized in the office of Hon. Chief Minister. It was decided that a committee should be formed under Chairmanship of Secretary PWD, Govt. of Delhi. This committee was to have members from Ministry of Road Transport Govt. of India, Delhi Development Authority, DUAC and PWD. The committee should invite proposals for these intersections from all interested individuals. These proposals should be examined by the committee and proposal approved for implementation. The DUAC to be informed about approved proposal for their record.

Accordingly, a committee was formed under the Chairmanship of Secretary PWD. A press advertisement was issued and it was notified that anybody could give proposals for concept design of these flyovers. There was good response and many proposals were received. Committee members examined these proposals on merit and finally three Consultants were selected for Dhaula Kuan flyover and three for AIIMS flyover. Consultants for selected proposals were requested to submit detailed drawings. After receipt of drawings the Committee examined and one design each for both the flyovers were selected. The Dhaula Kuan flyover was awarded to M/S SC Jain and Associates Architects and Planners whereas AIIMS flyover was awarded to M/S Mathur Kapre and Associates, Architect and Planner. For AIIMS crossing solution for all the sixteen movements was found with the help of two Clover-leaves. For Dhaula Kuan crossing, which was a five arm crossing, two level solutions was found using four clover leaves and slip roads for all the twenty movements. Both the locations required additional land beyond the available right of way.

AIIMS FLYOVER

AIIMS Flyover was to be constructed at intersection of Ring Road running East to West and Aurobindo Marg running North to South. For geometrics and structural design work, associate consultants were appointed. They were M/S Tandon Consultants Pvt. Limited for structural design and M/S CRAPHTS Infrastructures for traffic and transportation detailing. The preliminary plan of the proposed Flyover was prepared considering existing open space on North side of the intersection. It was noted that, additional land was required. The field survey was undertaken to appreciate suitability and requirement of additional land. The position on all the four sides was as follows: (i) On North East side, land of residential quarters and CPWD Enquiry Office Complex of East Kidwai Nagar, was to be taken over. Besides, there was a Rama Lila Ground & a Community Centre in this area, (ii) On North West side land of Govt. residential quarters of West Kidwai Nagar and a portion of Delhi Haat was to be taken over, (iii) On South East side, some land was required from AIIMS Complex and (iv) On South West side land was required from Safdarjung Hospital Complex. Besides, it was necessary to shift (a) One Petrol Pump; (b) Kashmir Migrants Market; (c) One Temple; (d) Taxi Stand and; (e) ten number of Kiosks.

The layout plan and levels were finalized to suit the site conditions. Loops were designed by keeping speed restricted to 30 km per hour. In order to restrict height of Flyover above average ground level, the Aurobindo Marg, (running North to South) was depressed by 2.5 meters and Ring Road (running East to West) was raised by 4.5 meters. In this area a number of services were running, which included water supply lines, electrical overhead and underground lines, sewer lines, storm water drains, telephone lines unfiltered water supply lines, Gas pipe lines etc. In the location plan these services were marked. It was necessary to relocate these services.

In order to get land and possession of structures to be demolished, respective authorities were approached and their approval taken. This included, Directorate of Estates under Ministry of Urban Development for residential quarters of East and West Kidwai Nagar as also CPWD Enquiry Office complex, Ministry of Health for Safdarjung Hospital and AIIMS Authorities. In the first stage, these buildings were to be got vacated and then taken over for demolition. Similarly, for shifting of services local bodies and other authorities were contacted and estimates for diversion of services were taken from them. These estimates were approved and payment made.

Separate corridors for shifting of services were planned. In all fourteen different service lines were to be shifted. About Rs. 15 crores were spent for shifting of services. Simultaneously diversion plans were made for the traffic to run on these important roads, with minimum possible inconvenience during construction stage. These Diversion Plans were sent to Traffic Police and their clearance obtained. Different plans were made for different stages of construction. For services to be diverted, plans for new alignment were made in consultation with respective agencies. The work of shifting of services was an important activity. Besides, as required by Department of Environment and Forest, PWD re-located some trees and re plantation was also to be done. In order to ensure that expeditious action is taken by different agencies for shifting of services, Coordination Meetings were held in the Offices of Chief Engineer and Secretary PWD. Shri Parvez Hashmi, Hon'ble PWD Minister, took interest and arranged meetings of all concerned departments, local bodies etc. to ensure expeditions action for shifting of services. In order that road users are not inconvenienced, a number of direction boards were provided at different places. Even temporary lights were provided for convenience of road users. The work of Metro Rail was not started at this location. Still the proposed alignment of Metro Rail was ascertained and considered for planning and design of foundations.

For speedy construction and to ensure proper quality, modern machinery and equipment were used which included Automatic and computerized concrete batching plants, paver for Rigid Pavement, mobile and stationary concerts pumps, WMM paver, bitumen paver, piling rigs, cranes, motor graders etc. The construction was done in different phases so that minimum disturbance was caused to road users. These phases were as follows: (i) First phase – Before diversion of traffic, piling work was started except along main Aurobindo Marg. Besides, two abutments and three piers were also constructed. The location of piles at the junction were suitably adjusted for proposed Metro line so as not to interfere with tunnel which was to come at a later stage. Location of piles was also intimated to Delhi Metro Rail Corporation (DMRC), (ii) Second phase – One Carriageway of Aurobindo Marg, towards AIIMS was locally diverted below already constructed span so that next span could be started, (iii) Third phase – The carriageway of Aurobindo Marg towards North was diverted below already completed span, (iv) Fourth phase – In order to construct the skew flyover, the Ring Road was diverted towards North. Thereafter pile driving as also casting of pile caps and piers was done. In some portion where traffic was yet to be diverted, only piling was done,(v) Fifth phase – The traffic was locally diverted on Aurobindo Marg, where piling was done but pile caps were not constructed and (vi) Sixth phase – In order to construct the underpass, the Aurobindo Marg was locally diverted and excavation done. Pedestrian Subways - Two pedestrian subways were constructed across Ring Road as a part of this project. For crossing Aurobindo Marg, pedestrian subways were already available.

Other related works – (i) Drainage and ground water recharging- The Flyover is spread in an area of more than 4 hectares with a number of loops and green spaces in between. The storm water drainage was designed to be gravity based, as it was the most economical and foolproof system. Trunk mains were laid along the periphery of the flyover, (ii) Road Signage -The interchange caters to 13 directional movements over two flyovers, one underpass and a set of loops running at different levels. In order to ensure effective guidance to the traffic using the interchange, a scheme of overhead and kerb mounted informative road signs were devised and the road signs were fixed in a planned manner,(iii) Lighting – Efficient lighting of the interchange area was done for safe movement of traffic and pedestrians. The design pattern demanded more attention, due to complicated configuration where individual carriageways must have light of adequate illumination level and uniformity. This has been provided by a judicious mix of 13 number of 30m high light masts with conventional street lights along slip roads in two quadrants. The landscaping was planned in such a way that no shadow reaches the carriageways even when plants grow up. Special care was taken in the areas below the flyovers. In case of failure of electric supply, these high masts and the pedestrian subways could be partially lit up through an emergency generator. A dedicated sub-station building was provided for these services,(iv) Planting of Trees and Horticulture work – A number of trees were affected by the interchange configuration. The trees were got inspected by the Forest department and as per their recommendations, a total of 147 trees were transplanted and 156 trees were cut. For every tree that was cut, 10 trees were planted at other locations as compensatory plantation. Between different loops and slip roads of flyover and between Aurobindo Marg and loops, the land was filled and earthen slopes were created. These sloping ground areas were visible to road users. Extensive horticulture work was done to give a pleasant view to road users. The plantation scheme was implemented in superb manner which resulted in extensive green look.

For conceptualization of initial planning of this project, initial survey work, identification of services, submitting proposals to different organizations regarding services etc, S/Shri HK Srivastava and SR Pandey, Superintending Engineer, along with their Executive Engineers S/Shri H R Tyagi, D C Goel and Pradeep Gupta did excellent work. The project was implemented under the leadership of Shri SPBanwait, Chief Engineer. Besides, Shri SS Mondal Superintending Engineer and Shri Rajeev Singhal did outstanding work. All the electrical works were done under the able supervision of Shri SC Khurana Superintending Engineer (Electrical). Horticulture wing also did outstanding work, Shri DC Azad was Deputy Directorand Shri Balbir Singh was the Assistant Director.

DHAULA KUAN FLYOVER

Dhaula Kuan location has historical importance and finds mention in Mughal history. Dhaula Kuan means white well and refers to an ancient water well in the area that had white sand. It was informed that this well was constructed by Shah Aalam II, the Mughal Emperor of India during 1761 – 1806 etc. period. At Dhaula Kuan, Ring Road Running from North West to South East, meets Sardar Patel Marg on North East, Ridge Road on North and NH– 8 on South West. It was busiest intersection of Delhi during eighties. The traffic at that time was exceeding 2 lakh Passenger Car Units (PCUs) equivalent per day. The basic concept of planning for this junction was conceptualized by M/s Mathur Kapre and Associates. The requirement was that traffic coming from all the five roads should be able to move to remaining four roads, signal free, with smooth curves and proper gradient. To start with the plan was made for a depressed clover-leaf with additional link for connecting Ridge Road. The land requirement as per detailed plan was finalized and land envelope marked at site. Thereafter, detailed survey work was done to plot the topography and make a contour plan. Besides, all services were marked on the plan. Diversion plans were also finalized for different stages of construction.

For this flyover land was required from different organizations. Most of the land was belonging to Ministry of Defence and some land was with DDA. Besides, Forest land was also to be taken over. One Police Station was to be shifted. Letters sent by PWD to Ministry of Defence Authorities did not bring fruitful results. The difficulty was brought to the kind notice of Shri VK Kapoor Hon'ble Lt. Governor of Delhi. He discussed the plan at length and critically examined it to ascertain as to whether requirement of land could be reduced. He appreciated PWD's requirement of land and difficulties in getting from Defense Authorities. It was agreed by him that matter would be taken up at highest level. Accordingly, a meeting was arranged with Hon'ble Defence Minister and our delegation led by Hon'ble Lt. Governor met him. Hon'ble Defence Minister agreed in principle, for handing over Defence land for construction of Flyover. This land was mostly in the compound of Defence Services Offices Institute. Besides, other concerned authorities were approached which include Forest Department, DDA and Delhi Police.

The Land decided for construction of flyover was having underground and over head services and an Electrical Substation. Estimates were asked from different organizations for shifting of services which included Mahanagar Telephone Nigam Limited for telephone cables, Gas Authority for Gas pipe line, Delhi Vidyut Board for HT lines/LT lines / Sub - station, NDMC, MCD, Delhi Jal Board and CPWD for other services like water supply line, sewer line, drains etc. All the Agencies were requested to plan relocating of services and submit their Estimates for making payment to them by PWD. The estimated cost of expenditure to be incurred by these agencies was included in Preliminary Estimate. The total Estimated Cost of project exceeded Rs. 100 Crores. It was first estimate in Govt. of Delhi that an estimate of cost exceeded Rs. 100 Crores. Therefore, the Estimate was submitted for Cabinet approval. In the cabinet meeting where Engineer-in-Chief and Chief engineer were present, the estimate was approved.

Implementation of Project – On the basis of estimates submitted, advance payments were made to different agencies for land and shifting of services. Tenders were invited, for the flyover work and M/S UP State Bridge Construction Corporation were lowest tender. Before award of work, it was necessary to ensure the availability of land. There was reluctance on the part of different agencies and the project was getting delayed. When agencies did not cooperate for expeditious action for shifting of services, matter was brought to kind notice of Government of Delhi. Several meetings were conducted in the Chamber of Shri Parvez Hashmi Hon'ble PWD Minister and senior level officers of different agencies were invited. Still there was considerable delay in the implementation of project on account of delay in shifting of services.

In spite of the fact that Hon'ble Lt. Governor met Hon'ble Defence Minister, the Defence Ministry was very reluctant to transfer the required land. In fact, their demand was to give them alternative land and construct their buildings. Only after that the Defence land would be made available. It was an unreasonable demand and not practicable. Similarly, Delhi Police did not shift its Police Post on the plea of VIP Security requirements. It looked that the project would not proceed and get abnormally delayed. PWD Engineers did the exercise of re-adjustment of plans, to fit into the available land. The revised plans were to meet requirements of MORTH standards. A number of alternatives were considered and a workable adjusted plan was finalized after the due process of approval by the Committee under the chairmanship of Secretary PWD. Engineers deserve complement for this revised plan, which was finally implemented. The terms and conditions of the Tender of UP State Bridge Construction Corporation were re- negotiated and thereafter the work was awarded to them. Thereafter the project was implemented in minimum possible time. Efficient supervision by PWD Engineers was highly appreciated. The completed project served its intended purpose and traffic conditions improved to a great extent at Dhaula Kuan intersection. The horticultural wing did excellent work and carefully planted trees and shrubs in the land around.

For this Project also S/Shri HK Srivastava, SR Pandey Superintending Engineer along with their Executive Engineers S/Shri H S Luthera, Khatwani and, V P Gupta, did outstanding work at the stage of conceptualization and initial planning. During construction stage the team was led by Shri BK Chugh Chief Engineer. For this project, Shri SC Khurana was Superintending Engineer (Electrical). The horticulture wing consisted of Shri Ganga Ram Dy. Director and Mrs. Veena Kantute Assistant Director. They all did outstanding work and their services were also highly appreciated.

LESSONS LEARNT

(i) Patronage of the Chief Minister was helpful in breaking the deadlock between PWD and DUAC. Later, help of the Defence Minister, and other high offices were taken to expedite availability of land and shifting of utilities. Project authorities must take help and advantage of high offices in resolving issues rather than delay the project.

(ii) At the planning stage efforts were made repeatedly to find a solution to the satisfaction of approving authorities. Persistent efforts of the team paid off.

(iii) All efforts were made to minimize inconvenience to road users by providing different traffic diversion plans at different stages of construction. Appropriate signages were provided for guidance on these diverted routes. Work operations were planned in advance and required equipment deployed for speedy construction.

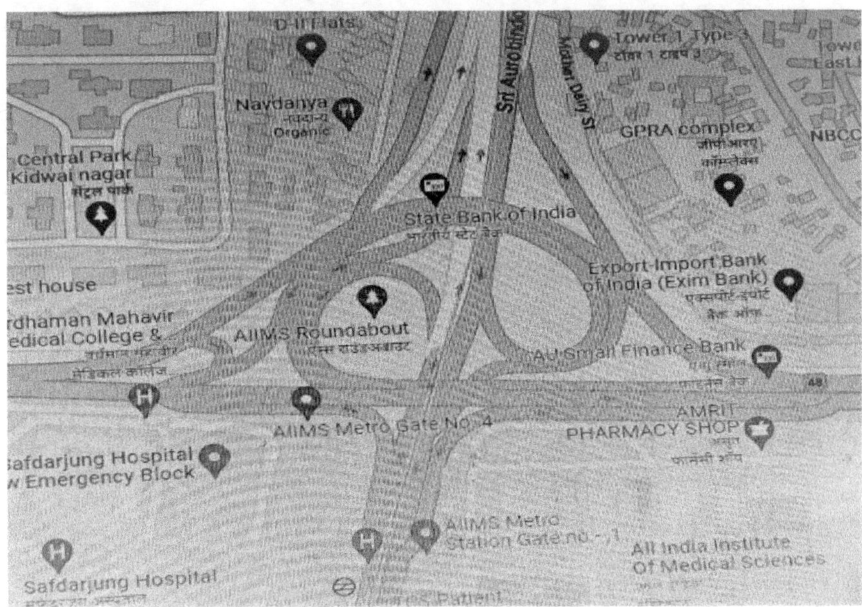

AIIMS and Dhaula Kuan Flyover Projects at Delhi -
Satellite picture of AIIMS crossing showing signal free traffic interchange

AIIMS and Dhaula Kuan Flyover Projects at Delhi -
Panoramic view of the AIIMS interchange. Aurobindo Marg is lowered to achieve
signal free traffic movement

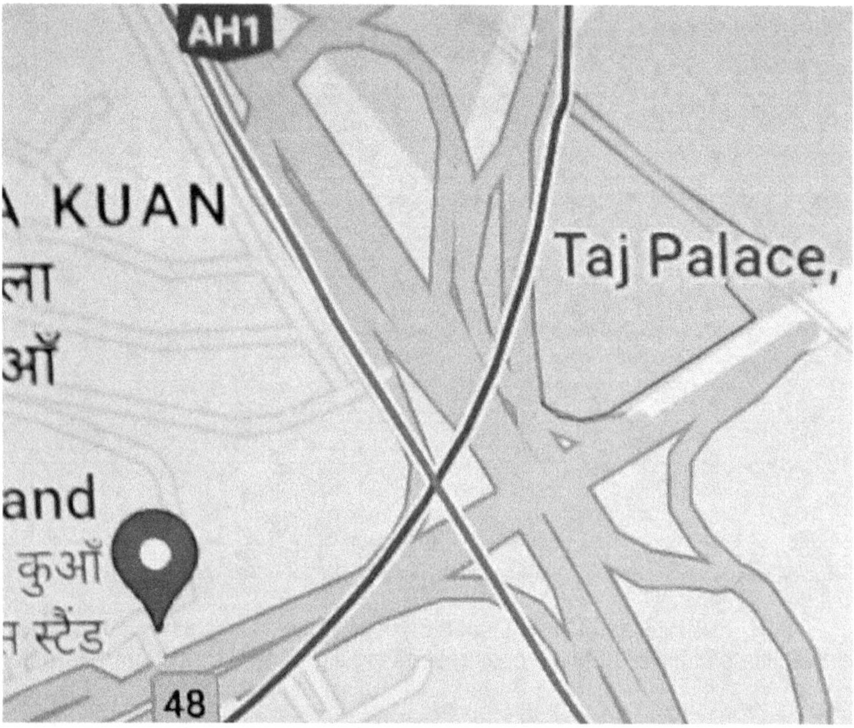

AIIMS and Dhaula Kuan Flyover Projects at Delhi -
Satellite picture of Dhaula Kuan Interchange

AIIMS and Dhaula Kuan Flyover Projects at Delhi – Dhaula Kuan Flyover Traffic on
Ring Road moving below Sardar Patel Marg flyover.

AIIMS and Dhaula Kuan Flyover Projects at Delhi – Dhaula Kuan Flyover Panoramic view of the Interchange. NH-8 – Sardar Patel Road is raised, Ring Road lowered. Top two corridors are Metro Pink Line and Airport Express Line constructed later.

AIIMS and Dhaula Kuan Flyover Projects at Delhi – Dhaula Kuan Flyover Landscaped area beautifully merging with concrete structures

FLYOVERS WITH PRECAST SEGMENTAL TECHNOLOGY ON RING ROAD AND OUTER RING ROAD AT DELHI

1. DECISION ABOUT CONSTRUCTING FLYOVERS

During 1995, a meeting was held in the office of Hon'ble Lt. Governor, to review the traffic situation on important roads of Delhi and for taking up construction of flyovers according to requirements. The meeting was attended by Shri KB Rajoria Chief Engineer and Planners from PWD Delhi as also DDA. The status of Ring Road and Outer Ring was discussed during the meeting. It was submitted by PWD Engineers that for corridor improvement on Ring Road and Outer Ring Road studies were conducted in the past to decide priority locations for construction of flyovers, based on the traffic volume and congestion. The locations identified on Ring Road included (i) Ashram Chowk, (ii) Khel Gaon Marg Crossing, (iii) All India Medial Institute Crossing, (iv) Africa Avenue Crossing, (v) Rao Tula Ram Road Crossing, (vi) Dhaula Kuan crossing, (vii) Kirti Nagar Crossing, (viii) Raja Garden Crossing, (ix) Rohtak Road Crossing, (x) Netaji Subhash Place crossing, (xi) Vehicular Subway at South Extension adjacent to AIIMS crossing. On Outer Ring Road, locations identified were (i) Kalkaji Road Crossing near Nehru Place and (ii) Savitri Cinema Greater Kailash 2 approach road. Besides, out of above locations, projects were already identified at some locations and preparation was already started by PWD Delhi and Delhi Transport and Tourism Development Corporation (DTTDC). Work was to be taken upon Ring Road at remaining locations. These were (i) Khel Gaon Marg Crossing, (ii) Africa Avenue Crossing, (iii) Rao Tula Ram Road Crossing, (iv) Kirti Nagar crossing, and (v) Netaji Subhash Place crossing. On Outer Ring Road, work was to be taken up at for both the locations that is (i) Nehru Place and (ii) Savitri Cinema. It was decided during the meeting that except for location near Netaji Subhash Place crossing, all other flyovers will be constructed by PWD. Out of these, PWD was directed to take up in the first instance two flyovers at Ring Road, one on Rao Tula Ram Marg (RTR) Crossing and another on Africa Avenue crossing. On Outer Ring Road both the flyovers at (i) Nehru Place and (ii) Savitri Cinema were to be taken up.

2. STATUTORY APPROVAL BY THE EMPOWERED COMMITTEE

During the meeting, issue of delay due to approvals by Government agencies before a flyover project could be taken up also came up for discussions. It was decided to constitute an Empowered Committee to facilitate and expedite clearances so that there is no further delay. The Empowered Committee was accordingly formed by Govt. of Delhi. The Secretary PWD was to be the Chairman of the Committee and Members were Chief Engineer (T&T) & Chief Engineer (Bridges) of Ministry of Road Transport & Highways, Commissioner (Planning) DDA, Secretary DUAC, Adviser DUAC, Chief Engineer CPWD, Senior Architect PWD and Chief Engineer, PWD. The design selected by the Committee was deemed to have acceptance/approval of the departments/organization represented in the Committee. Thus, after the completion of the task, the Committee was to send approved plans for record and acceptance by these departments/organizations.

3. PACKAGES FOR DESIGNS AND CONSTRUCTION

As regards manner of taking flyover projects for construction, there were divergent views. These projects could be constructed by awarding design and construction in one package or first the design work could be awarded and later on construction work. Hon'ble Lt. Governor wanted implementation of projects expeditiously. PWD decided that it would first tender to get the design work done. Later, on the basis of these designs, tender for construction work will be called. It was expected that with this approach, the project would be economical and faster. Besides, uncertainties would be avoided. Preliminary Estimates submitted by PWD for these four flyovers were approved by Government, without any delay.

4. PRELIMINARY INSPECTION

(i) **Ring Road** - Both RTR Marg Crossing and Africa Avenue crossing were having very heavy traffic. Besides RTR Marg was frequently used for VVIP movement. Thus, during construction, free traffic movement was required both on Ring Road and cross roads. Besides, along the Ring Road water supply mains, sewer lines, Electrical overhead lines were located. Some of these services were required to be shifted prior to start of flyover construction activity.

(ii) Near Outer Ring Road tri-junction with Road leading to Greater Kailash Enclave 2, the available road width was limited. Besides water supply mains, sewer line and out fall for main drain was also existing at this location. On Outer Ring Road intersection with road to Kalkaji near Nehru Place, there was heavy traffic and existence of bulk service lines. Near these locations there was hardly any space for locating casting yard.

5. GEOMETRIC DESIGN

The traffic studies and geometric design of these flyovers was got done from M/s CRAPHTS represented by Shri D Sanyal, a top level expert in this field. Both Ring Road and Outer Ring were six lane divided carriageway. Thus, the existing width of road was 11.5m dual carriageway. For Flyovers, it was decided to provide 9m wide dual carriageway. To decide height of flyover, the clearance from road to bottom of flyover was taken as 5 meter. The gradient of approach roads was kept as 1 in 30 but in exceptional circumstances at some location, 1 in 25 gradient could be allowed. Plan and L-Sections with these geometric designs were prepared by consultants and after detailed examination at respective project sites, these were approved for implementation.

6. FOUNDATION DESIGN

After subsoil investigations, the foundation design was finalized. It was decided that one meter diameter bored compaction piles would be provided. For these piles, M-35 Grade concrete with Portland slag cement was recommended by the consultant. RCC piers with external molded design were approved. The purpose for this design was to discourage putting advertisement posters on piers. The pier size was kept as 1.5m x 2m. For piers, M-35 grade concrete was to be used.

7. DESIGN COMPETITION FOR FLYOVER

It was decided that design proposals should be invited from structural consultants and should be examined on merit. A press advertisement was accordingly given. To examine and approve proposal a committee was formed under the Chairmanship of Shri Rakesh Mohan, Secretary (PWD). Engineer-in-Chief, Concerned Chief Engineers and Superintending Engineers were members of this committee. A number of design proposals were received from different consultants and these were examined by the committee. It was general opinion that the proposal with minimum construction activity at project site should be preferred because no open space was available near these intersections. After examining different designs, the proposal given by M/s Mahesh Tandan and Associates, with pre-cast segmental construction was approved. This technology was earlier used for elevated road construction in other countries. It was to be tried for the first time in our country. It was expected that dislocation of existing services would be minimized by adopting this technology. The construction time was also likely to be reduced. As per this proposal, the pre-casting of members was to be done at a casting yard, which could be located at a suitable place, for all the flyover projects. Besides precast construction was to be done under controlled conditions. Therefore, it was expected to give better quality, higher strength and in turn more durability.

8. APPROACH RAMPS

For approach ramps of flyovers, in past RCC Retaining walls were provided. An alternative was considered, which was economical and faster. It was Reinforced Earth technology, which was comparatively new in our country. Shri KB Uppal of M/s Aimil, introduced this technology to PWD. Besides a presentation was given by representative of the Reinforced Earth technology expert from France. The technology was based on balancing active pressure and passive pressure on precast concrete panels, to be used for retaining earth on approach ramps. On these precast panels special reinforcement bars were to be hooked. These bars were to be embedded in the retained earth to provide passive resistance. Thus, there was stability of structure. It was economical compared to RCC retaining walls. Besides, better quality control was possible by using pre-cast panels, manufactured at casting yard. This alternative was approved for these flyovers.

9. INTRODUCING CONTROLS AS PER ISO 9000 STANDARDS

To achieve quality standards of concrete work in the project, the procedure recommended in Indian Roads Congress publication, IRC-SP 47 was followed. In fact, this document was based on ISO 9000 standards. It was drafted by eminent Engineers including Shri NV Mirani, Secretary PWD, Govt. of Maharashtra, and Shri SV Reddy from Gammon India, a well known authority on RCC work. For the first time in the country, these standards were followed. In order to ensure proper implementation, for controlling process, PWD Engineers were given training in ISO standards and certificates were taken from ISO system registration agencies. Thus, absolute control on process was exercised to assure quality.

10. GEOMETRY OF FLYOVERS AND CASTING OF BEAMS

The span of main flyover in the middle was kept as 41.1 meters and two adjacent spans were 31.05meters. The end spans were 26.1 meters. All the beams were cast at Nizamuddin Bridge Casting Yard, with design mix under controlled conditions. Segments were match casted, using long line method of centralized casting. The segments were of two types, with uniform depth and with variable depth. M-45 Grade concrete was used. Proper curing was done and for stacking proper system was followed. These segments were cast as per specific requirements at the site.

11. ERECTION OF BEAMS

Erection of beams was done by providing two sub bridges per carriageway and then joining by cast in situ stitches. The erection of two span assembly girders was done by using triangular trusses with hangers, to hold these girders during erection. All staging were separated by 1.8 meter vide gap. There were three such gaps which were used for placing trestles under the assembly and for locating jacks, for permanent placing, while erecting. Bearings were placed and grouted under each support location, when work up to that stage was completed. After erecting all the four stages, the central stitch was cast. The cable installation was done continuously. In brief the erection steps were as follows; (a) Placing and positioning of trestles over concrete pad, (b) Adjusting level of top cross beam of trestle with hydraulic jack, placed in extended ram condition, (c) Lifting and placing each of two triangular trusses on top of trestle, (d) Two trustless were locked to prevent any relative movement, (e) Crib support around the pier was erected. Step one for piers P2 or P5 and steps two and three for piers P3 or P4, (f) Bearings were placed in position over leveling screws, (g) Lifting of the pier segment was done. Hangers were providing for supporting the segment with turn-buckle. Erection was gradually done. First the temporary prestressing was done. After all members were placed at their final position, permanent prestressing was done.

12. CONSTRUCTING RETAINING STRUCTURE

Approach ramps were constructed using Reinforced Earth technique. The structural principle was very simple. Outside walls of precast concrete panels, were interconnected. Some of these panels were tied with high adherence (ribbca) steel strips, termed as soil reinforcement. The filling of ramps was done with Jamuna sand and compacted. To achieve maximum density, the filling was done in optimum moisture content. Besides, proper drainage arrangements were done under the earth work. Precast panels were retaining the earth and it was exerting pressures on outside. The reinforcement was retaining panels on account of resistance of earth on reinforcement. Thus, it became a stable structure. Testing of reinforcement was done by pullout tests. Crash barriers were provided to avoid severe accidents on the flyover.

13. TIME FOR CONSTRUCTION

The time taken for construction of four flyovers was about one and half years. It was the fastest construction ever achieved anywhere the country. Besides, highest level of quality standards was achieved.

These projects were executed when Shri KB Rajoria was E-in-C, under the leadership of Shri S P Banwait Chief engineer. Shri NMD Jain Superintending Engineer and S/Shri Shishir Bansal, RC Rangre and Tarkeshwar Tiwari were Executive Engineers. They did outstanding work.

LESSONS LEARNT

(i) **Delay analysis and decision making:** Many a times it becomes necessary to analyze the system and procedures if the existing ones do not provide desired outcomes. To overcome procedural delays in this case, system of approval by duly constituted Board was adopted. Delays were thus avoided and Engineers could focus on main work.

(ii) **Adoption of New Technology and Risk Management:** Even when, technology is already proven and put to use in other countries, applying the same in our country is a courageous initiative and some risk is involved. It was the confidence on the work capability of the team that precast segmental flyovers were tried for the first time in India. This technology was selected, because there was limited space around these locations and these was difficulty for storage of materials and also for diversion of traffic during construction. Precast segments were manufactured at a casting yard near Nizamuddin Bridge where space for manufacturing and storage of precast elements was available.

(iii) **Initiative for Cost saving and quality improvement:** By constructing flyovers with this technology, the shifting of underground and overhead services was avoided to a great extent. Thus, there was saving of expenditure and time taken for construction of flyover was also reduced. Precast construction had better quality control and in turn increased durability.

Flyovers with Precast Segmental Technology at on Ring Road and Outer Ring Road at Delhi - Variable depth precast segments were provided in central and adjoining spans for aesthetics and economical design.

Flyovers with Precast Segmental Technology at on Ring Road and Outer Ring Road at Delhi - Elegant precast concrete panel for reinforced soil retaining structure in ramp

Flyovers with Precast Segmental Technology at on Ring Road and Outer Ring Road at Delhi - Use of Molded design of RCC piers for aesthetics

26

FLYOVERS WITH STEEL FABRICATION TECHNOLOGY ON RING ROAD AT DELHI

1. FLYOVERS WITH STEEL FABRICATION TECHNOLOGY ON RING ROAD AT DELHI

After award of work for four flyovers on Ring Road and Outer Ring Road, with Precast Segmental Technology, two more identified locations on Ring Road were, (i) Kirti Nagar Crossing and (ii)August Kranti Marg Crossing. At both these locations the traffic was exceeding 12 thousand PCUs per hour. Government gave approval to construct flyovers at these locations. Thereafter, by appointing a traffic engineering consultant for both these locations, traffic and land survey work was undertaken. Besides, the consultant was also requested to give geometric design. Brief details for these locations are as follows: (i) Andrews Ganj Intersection- At this location the Ring Road intersects with August Kranti Marg. It is near New Delhi South Extension Market. The August Kranti Marg connects New Delhi towards North side and Asiad Villages on the South. There was large pedestrian traffic and a number of vehicles were taking U-turn at this location. Besides, in the vicinity several minor Roads of different colonies were joining Ring Road. (ii) Mayapuri Intersection - This intersection is at North Western approach of existing Rail Over- Bridge on Ring Road at Naraina. On account of Industrial and Residential areas in the vicinity, it is heavily congested location.

2. GEOMETRIC DESIGN AND SELECTION OF TECHNOLOGY

After study of geometric survey plan it was noted that at both these locations, Ring Road was curved and not straight. Therefore, the Flyovers were to be Skew and not at right angle. Besides, for convenience of traffic the Flyovers were to be constructed along Ring Road. It was decided that the obligatory central span should be kept 45m. At Andrews Ganj, two approach spans of 26 meters each were proposed on either side of main flyover. At Mayapuri, the Naraina side approach was to be merged with existing ROB, to avoid roller coaster effect. To keep the viaduct on Nariana side horizontal, approach spans were provided towards Naraina side. On account of non availability of land, the width of carriageway was kept limited to 9.5 meters compared to standard practices of providing 11.5 meters width. The clear height available for traffic was kept 5 meters, according to applicable norms. Most important parameters considered for selection of technology were, (i) Speedy construction, (ii) Minimum activity at project site, (iii) Minimum disturbance to traffic, (iv) Minimum relocation of underground services, (v) Cost effective solution, (vi) Latest technology, (vii) Aesthetics and (viii) Durability. For selection and approval of consultant, the committee under the Chairmanship of Secretary (PWD) Govt. of Delhi deliberated at length. Shri KB Rajoria Engineer-in-Chief, Shri SP Banwait Chief Engineer and Superintending Engineer In-Charge were members of this Committee. It was decided that consultants should be requested to give proposals of different technologies and designs. It was also decided that the selected consultant would be paid 1% of the cost of project. Several proposals were received, which included conventional post tensioned beams, segmental construction, steel structure, composite structure etc. Out of these proposals, the steel, concrete composite design was selected on account of specific boundary conditions. This proposal was given by M/s CES, a renowned consultancy company. They took services of Shri Shitla Saran retired Chairman and Managing Director of UP State Bridge Corporation (UPSBC). He was well known Bridge Engineer of the country and construction of a number of bridges was to his credit. At that time, Steel concrete composite design was not common in the country and no project was implemented in the country with this technology. Of course, Shri SS Chakraborty CMD of M/S CES mentioned about proposed for steel flyover project, accepted by Govt. of West Bengal authorities at Kolkata.

3. VISIT TO FAMOUS WORKSHOPS FOR STEEL FABRICATION

In order to understand the steel fabrication technology as also to find feasibility and capability, it was decided to visit prominent Workshops in the country. Top level Engineers of PWD visited Kolkata, Mumbai and Chennai where these workshops were located. At Kolkata, details of the project mentioned by the Consultant were ascertained and it was found to be at preliminary stage. So, no details could be given by authorities. Thereafter, Engineers contacted M/s Bridge and Roof, a famous bridge construction company. They mentioned that their practice was to undertake fabrication at the project site and there was no workshop of their company. Therefore, their quality parameters could not be ascertained. Further the workshop of M/s BBJ, a Govt. of India Undertaking was visited and it was found appropriate and proper for undertaking such jobs. Thereafter, the team went to Mumbai and visited the workshop of M/s Mazgaon Docks another Govt. of India Undertaking. This Workshop was also found appropriate and suitable. Finally, PWD team went to Chennai and the workshop of M/s L&T was visited and it was also found suitable. No other workshop of equitable level could be found anywhere in the country. After deliberation, with Engineers it was decided that in order to ensure quality of highest standard, the fabrication of steel work should be done at one of the three workshops and welding under controlled conditions should be allowed. Besides, it was also decided that the welding work should not to be done at project site and as per requirement bolting/riveting should be done to join workshop fabricated components. For substructure both steel and concrete options were considered suitable.

4. SELECTION OF PROOF CONSULTANTS

The Steel Flyover proposed by the Consultant was first of its kind in the country. For proof checking both CPWD and PWD Delhi did not have structural designer with required experience. Therefore, a proof consultant was necessary. Shri TN Subbarao who was heading M/s Construma was selected as proof consultant. Shri Subbarao was a famous structural engineer in Highway sector. The selection was on clear mandate that he would personally check the work done by the Consultants. In fact, there could not have been a better selection for proof consultancy in the country.

5. AWARD OF WORK

Detailed designs were prepared by consultants. Thereafter, estimates were framed and tenders called. In the tender document, a special condition was given mentioning that fabrication of structural steel work should be done at one of the three workshops short listed by PWD. These were (i) BBJ Kolkata, (ii) Mazgaon Docks, Mumbai and (iii) L & T Chennai. Contractors were short listed on the basis of their experience and financial capabilities. Tenders were called only from these agencies. M/s AFCONS were the lowest tenderer and after due scrutiny, the project was awarded to them. After award of project, they did not start the work for quite some time. Instead, their representative met PWD Engineers and requested that they should be allowed to do steel fabrication at project site. It was not agreed. It appears that some influential persons at higher level were approached by contractors. They in turn advised PWD to consider request of contractor. The representative of firm again met and repeated their request. They were informed squarely that their proposal to undertake steel work at project site, could not be agreed to account of the fact that it would be considered favor to contractors by change in contract condition. Besides, PWD was interested in quality work. They were given two options, either to undertake the work as per tender condition or withdraw their tenders. Sometime was given to them to take a decision. Finally, their representatives met Engineer-in-Chief and informed that their management took decision to undertake project as per conditions of contract and complete it.

6. PRELIMINARIES AT PROJECT SITE

According to General Arrangement Drawings, additional land beyond right-of-way was not required. But numbers of services were to be shifted for taking up the work. Besides, spaces were also to be identified for placing erection cranes for girders. At the site, soil investigations were done for foundation design. Besides, the location of services was ascertained. It was noted that available right-of-way near project site was only 55 meters as against 66 meter right of way of Ring Road. In this available width of land, a diversion of road during construction stage was planned, so that there would be minimum disturbance to traffic movement during implementation. As regards services, it was decided to consider relocation, by accommodating in the right-of- way. Services relocation was coordinated with different agencies like Jal Board, Electricity authority etc.

7. GENERAL STRUCTURAL ARRANGEMENT

Pile foundation was considered suitable and 1.2 meter diameter and 26 meter deep, bored cast-in-situ RCC piles were provided. Below each pier one pile was provided. Two piles were interconnected. RCC piers were provided with groves on aesthetic considerations. The super structure was of built up Plate Girders, 1.1 meter deep for approach spans and 1.7m deep for central span. The structural steel was confirming to Fe- 540 as per IS: 8500. For fabrication of these girders, the contractor was directed to decide for one of the approved workshops. In turn they decided to engage L&T workshop at Chennai. While fabricating these girders, the camber as required was provided. The welding was done as per American Code, as no suitable Indian code was available. The engineers of PWD checked all the structural members at the workshop and only after approval by them, these were transported to project site. The erection work was done by experts from L&T. The components were bolted together with high strength friction bolts. This arrangement ensured that there would not be reversal of stresses. Epoxy based painting was done on structural steel, to last at least fifteen years. Elastomeric bearings were used. Cast-in-situ deck slab was provided. The approaches were made with reinforced earth walls.

8. QUALITY MANAGEMENT

These flyovers were important structures and new materials like steel was used. It was therefore decided to have Q-4 level of quality assurance as per provision of IRC SP-47(1998), and QA Assurance Manual was prepared. PWD, Delhi was the first agency in the country to follow SP-47. An elaborate Quality Assurance Plan was drafted and applied, for fabrication work at workshop. In spite of the fact that the workshop had ISO system certification, third party inspection was got done through Regional DGS&D unit, as an extra precaution.

9. The credit goes to officers who worked day and night for implementation of project. Both Contractors and Consultants deserve appreciation. For this project Shri S P Banwait Chief Engineer worked with full zeal and confidence. His contribution is highly appreciated. A special mention is made of S/Shri S S Mondal, Project Manager, Shri Rajeev Singhal Executive Engineer and Shri S C Khurana, Superintending Engineer (Electrical). The team did outstanding work for this project under the guidance of Shri K B Rajoria, E-in C, Delhi PWD.

LESSONS LEARNT

(i) **Introduction of New Technology:** Construction of flyovers with steel fabricated beams and according to IRC SP-47, clearly demonstrated initiative, zeal and hard work of CPWD Engineers. It was the first project with this technology in the country.

Adoption of this method considerably reduced the time of construction as well as inconvenience to road users during construction.

(ii) **Quality Assurance in Construction:** To ensure reliability of quality of construction, a bold decision was taken to fabricate components in one of the top fabrication workshops of the country. Only three workshops could be shortlisted. The fabrication under controlled conditions using skilled and certified workmen, total checking by outside agency and cartage to site ensured that quality of highest level was achieved. For integrity of the quality of joints, no welding was permitted at project site. Fabricated members were joined at work site by bolts only. It had no parallel or equitable structure in the country.

The team firmly believed in the envisaged process of construction to achieve its objective. Even though there was undue pressure from different quarters to change the methodology of fabrication, PWD did not agree in the interest of ensuring quality. Courage of conviction is essential in such situations.

(iii) **Risk Management:** Acceptable technical solutions were found and incorporated in the design at the planning stage to overcome adverse situation arising out of typical location, availability of land etc. One has to be prepared to resolve such issues while working in congested urban area.

(iv) **Maintenance Cost:** Practically no maintenance cost on protective painting was expected for steel flyovers, for another fifteen to twenty years due to use of epoxy paint. It is appropriate to consider life cycle cost while taking decision about materials and technology.

Flyovers with Steel Fabrication Technology on Ring Road at Delhi -
45 m Central span with box girders at Mayapuri Intersection
Smooth right turning traffic

Flyovers with Steel Fabrication Technology on Ring Road at Delhi -
View of steel flyover from below

Flyovers with Steel Fabrication Technology on Ring Road at Delhi -
View from slip road towards flyover

27

USE OF FLY ASH FOR APPROACHES TO NIZAMUDDIN BRIDGE, DELHI

1. FLY ASH FOR EMBANKMENTS

Civil Engineers are fore-runner in developmental activities. They have a stake and a great responsibility towards conservation of environment. There has to be insistence on "Reduce", "Reuse" or "Recycle". During nineties, for the first time in the country, PWD Delhi took initiative to use fly ash, on large scale for approach road on high embankment to the new Nizamuddin Bridge at Delhi, constructed under Japanese Aid programme.

2. BACKGROUND

For the increased demand of power, coal was the major source of energy. Many Thermal Power Plants were setup in our country. The thermal grade Indian coal contains 35 to 45% of ash resulting in generation of huge quantity of fly ash. Fly ash is a dusty nuisance by-product of Coal based power generating plants. The ash collected from air pollution control equipment of the power plants are either transported in the form of slurry to the Ash Pond or deposited in dry form, in ash Mounds. Management and disposal of fly ash, treated as an environmentally hazardous waste material, has been an environmental issue. Storage of ash in ponds and mounds also require large areas of land. Fine particles carried away with wind causes air pollution. Fine particles flowing into rivers cause problem for aquatic animals. Thus, dust and erosion control measures are necessary for its storage and use. Potential of fly ash as an alternate engineering material remained untapped. From 1930 onwards, in USA, it was used in many projects as a structural backfill material behind retaining walls and bridge abutments. During 1950, use of Fly ash as a structural fill or embankment material was tried in Great Britain. Its relatively low unit weight made it well suited for placing over soft or low bearing capacity soils. Besides, its high shear strength compared with its unit weight resulted in good bearing support and minimal settlement. However, till nineties, it was an untried method for embankment construction in our country.

3. SECOND BRIDGE AT NIZAMUDDIN DELHI

In 1998, a new bridge at Nizamuddin, Delhi, was constructed on river Yamuna downstream of existing bridge, under Grant-in-Aid programme from Govt. of Japan. This was planned as a "Replacement Bridge", because existing bridge, constructed during early sixties, had developed symptoms of distress. The scope of the Grant-in-Aid from Japan included construction of a new bridge, with short approaches on either side, to connect to approaches of existing bridge which were to be dismantled. Later on, it was decided, in consultation with the Ministry of Road Surface Transport, Govt. of India, that to cater for traffic of 100,000 PCUs per day, earlier constructed distressed bridge, after retrofitting, would also be retained. Therefore, it was further decided that new approaches, would be constructed up to eastern and western banks of river, for connecting new Nizamuddin Bridge, which was four lane (14.5m) wide with 3.5 meters wide cycle track. For the new approach 8 meter high embankment was required, for connecting to banks of river. In consultation with Central Road Research Institute, it was decided to use fly ash for approach embankments because in Delhi Fly ash was available in abundance. This proposal was submitted to Ministry of Surface Transport and Highways for their approval as it was part of National Highway. They technically examined the proposal based on fly ash filling of embankment and did not approve the proposal for this project. Main reason put forward by the Ministry was that for such an important road link on National Highways, the embankment with new material with no earlier experience was not considered suitable. Thus, in a way the Ministry discouraged for use of new material (fly ash) inspite of the fact that, comprehensive technical study was conducted in collaboration with the Central Road Research Institute.

4. REFERENCE TO CABINET SECRETARY

In view of the fact that there was difference of opinion between Ministry of Road Transport & Highways and Govt. of Delhi, as per procedure, the matter was referred to the Cabinet Secretary. The Ministry of Urban Development was also involved for resolving the issue. A meeting was held in the Chamber of Cabinet Secretary which was attended by high level officers of both the Ministries and Govt. of Delhi. The Ministry of Road Transport and Highways was represented by Shri AD Narain, Director General. The PWD was represented by KB Rajoria as Engineer-in-Chief, Shri KN Agarwal Chief Engineer and Shri Anant Ram Superintending Engineer. The subject matter was explained to Cabinet Secretary and both sides gave their point of view. He enquired about importance of road and was told that it was the most important link between Delhi and East Delhi as also Noida and beyond. He wanted to know, if the embankment would get washed away due to rains, floods or otherwise, whether an alternative link was available. He was informed that existing approach of old bridge was available. Besides, there were other bridges on the river Yamuna in nearby vicinity. After careful consideration, it was decided that new technology of using fly ash should be tried, in national interest and proposal of PWD was approved. The project was implemented by using fly ash.

5. PRELIMINARY DESIGNS OF EMBANKMENT WITH FLY ASH

In the alignment of approach road, the subsoil strata predominantly consisted of non plastic silt and silty sand. It was considered as cohesionless material for shear and settlement considerations. In consultation with CRRI, the preliminary design of embankment was developed. It was decided to provide fly ash core protected by earth cover on exposed side (downstream) as well as on top. On upstream side, it was adjacent to embankment of existing bridge and, not exposed to river water. Materials to be used were, (a) Fly ash 1,30,000 cum, (b) Earth 1,05,000 cum and (c) Red Bajri 20,000 cum. To use fly ash in such a large quantity was a bold decision. Fortunately, Delhi PWD Engineers had courage for taking such decisions, on account of the fact that under confined conditions, fly ash was already used by PWD Delhi for Visvesvaraya Setu (ROB 22, Mathura Road) and Hanuman Setu (Monkey Bridge near ISBT) Flyovers.

6. SUITABILITY AS EMBANKMENT

The technical parameters of available fly ash were favorable, as it had low specific gravity. The main properties of fly ash were, (a) Grain size distribution- Fine grained material, (b) Atterberg Limits-Non plastic, (c) Specific gravity ranged from 1.2 to 1.4 as compared to soil between 1.65 to 2, (d) Compaction Characteristics were better than sand, (e) Permeability was high, (f) Shear strength was higher than normal earth, (g) It had low compressibility, (h) The predominant components were silicon, aluminum and iron in the form of oxides and, (i) Carbon was also present which acted as dilutant of active pozzolanic mixture in fly ash. Economic viability was considered by way of comparison with cost of conventional fill material that is earth. Most of the earth was to be transported from long distance and transportation cost was high. The indirect cost factor was safe disposal of fly ash on account of clearing land used for collection of fly ash. There was a direct saving to Delhi Vidyut Board, by way of shorter distance for disposal of fly ash. Besides, gainful utilization of fly ash was good for reduction of the environmental pollution.

7. SPECIFICATIONS FOR FLY ASH FILLING

The Specifications for fly ash embankment work were decided in consultation with the CRRI. Salient features of the specification being,(i) Clearing and Grubbing of site, (ii) Places where foundation of embankment was in an area with stagnant water, it was decided to dewater and ground supporting the embankment to be leveled and compacted. (iii) Fly ash and cover material (for erosion control) were spread in layers of uniform thickness, not exceeding 200 mm and compacted by mechanical means. Moisture content was controlled. For compaction, both vibratory and static rollers were to be deployed and 95% of optimum density was achieved. (iv) Same procedure repeated for each layer, (v) After completion the embankment from bottom to top had layers as follows – (a) 2 m fly ash (b) Soil cover (c) 2 m fly ash (d) soil cover (e) fly ash as per requirement of height, up to 2m, (f) Soil cover (g) Red Bajari and (h) Road crust

(vi) On the side slope one meter soil cover was provided. It was then covered by stone pitching for erosion control, (vii) After undertaking slope stability analysis, the side slopes of the embankment was decided.

8. IRC SPECIFICATIONS FOR USE OF FLY ASH -

After the successful use of fly ash on this project, as structural fill material, a report was sent to Indian Roads Congress. IRC recognized the need to coding the specification for use of fly ash on road work. The Geotechnical committee of IRC took a look at the Specifications followed for this project and drafted the specifications for further use of fly ash in road projects.

LESSONS LEARNT

(i) **Risk & Opportunity based decision making:** Use of about 1.2 lakh cum of fly ash in the approach road to the newly constructed bridge over river Jamuna was a path breaking application of engineering skills by Delhi PWD engineers.

With the help of CRRI, quality of the fly ash available within economical lead was studied, analyzed and process of construction was decided. Thus, It utilized huge quantity of environmentally hazardous material, reduced cost and paved way for drafting specifications for use of fly ash in embankment construction for the country.

(ii) **Proposal with Self-Conviction:** One should have courage to stand by one's conviction. Once the PWD Engineers were convinced about safety of the embankment under all possible adverse eventuality, they did not hesitate to take a stand.

(iii) **Action to address risks and opportunities:** Risk analysis was done. In case of any adverse eventuality, alternative routes were available.

1. NEED OF THE PROJECT

The population of Port Blair, the Capital of Andaman & Nicobar (A&N) Islands was assessed as per 1991 census as 75,000. However nearly 80,000 tourists visited these islands during the same year indicating tremendous tourism potential. The only means of transportation to these islands (from the mainland) is either through sea journey or by air. From Kolkata or Chennai, a Ship takes nearly 2 to 3 days to reach Port Blair. During rains, the sea often becomes rough and shipping services get suspended. In such circumstances air travel remains the only mode of travel to meet the travel requirements and medical emergencies etc. In short, undisturbed air connectivity to these islands is critical. The airport at Port Blair is a Defence airport managed by the Indian Navy. The orientation of the runway is 04/22 with length of 6000 feet. During early nineties, the runway pavement strength was assessed as PCN 18 F/C/X/U. According to the International Civil Aviation Organization (ICAO) standards, PCN defines the strength of the runway pavement and thus is an indicator of the load carrying capacity of a pavement. Because of steep hills surrounding the runway, the existing approach slope was much steeper than 1 in 30 allowed. There was also a steep hill close to 22 end, making the runway unidirectional thereby shifting the threshold by another 823 feet leaving effective runway length of only 5,177 feet. Thus, the runway was suitable for handling up to Boeing 737 type aircrafts only which were operating with load penalty, i.e., operating with reduced pay load compared to the aircraft capacity. Being close to the sea, the existing runway was liable to flooding and there were instances when flights had to be suspended because of flooding of runway. During mid-nineties, only 9 flights per week were being operated by Indian Airlines. Thus, there was a need to extend and strengthen the runway to enhance the safety of operations, be more useful to defence in future and also be able to handle more flights including chartered flights for foreign tourists. Accordingly, proposals/estimates were made by M/s RITES Ltd and an 'Expenditure Sanction' amounting to Rs. 49.29 Crore was accorded in the year 1995 by the Ministry of Civil Aviation, GOI. This sanction was to strengthen the existing runway pavement to PCN 34 from the existing PCN 18 and to extend the existing runway by additional 5,000 feet.

2. STATUS OF PROJECT

(i) Shri HS Dogra was posted as Chief Engineer and Secretary PWD in the Andaman and Nicobar Islands in August 1997 on deputation from CPWD. At that time the runway project was under progress for extension of the existing 1800 m runway to a total length of 3300 m and strengthening of the existing runway. The work of earthwork excavation in the extended portion of runway had been awarded a few months back but the work was almost at a standstill. The contractor had raised a dispute regarding classification of soil which was stated to be pending with the department. Apparently, the contractor had mobilized enough excavators and dumpers etc. for carrying out the earth work including its disposal to the designated dumping site.

(ii) After discussions with the Andaman PWD engineers, it transpired that as per contract a total of 37 lac cubic meter of earth had to be excavated. Simultaneously, about 9 lac cubic meter of good earth was to be filled up in the initial length of the extended portion of the runway. M/S RITES had prepared the Detailed Project Report and classified the total excavated soil as 94% all kinds of soil, 2% hard soil and 2% each as ordinary rock and hard rock. They had proposed to use part of the good, excavated earth for filling. But by August 1997, all good topsoil had been excavated and dumped in the dumping area. Thus, good soil for filling for the initial portion of the runway being extended was not available. The maximum depth of cutting was more than 10 meters. However, after about 2 meters of excavation, rock strata were found. The contractor argued that he had not organised enough equipment to handle such huge quantity of ordinary rock and hard rock, which was at large variance with the Bill of Quantities. Thus, he demanded that the rates for such large deviation in quantities of ordinary/hard rock be settled so that he could mobilize additional equipment like rock breakers etc. to proceed with the work.

(iii) M/S RITES Ltd were asked to clarify as to on what basis the soil classification was decided in the DPR. Their replies were vague stating that classification was done after making local enquiries. It was argued that the soil classification cannot be predicted at the depth of eight to ten meters by local enquiries. However, M/S RITES kept on giving such vague answers and went to the extent of mis-informing the A&N administration that the engineers of Andaman PWD were wrongly classifying the soil and even suggested to the administration to appoint them as Project Management Consultant to handle the project. After Shri Dogra took over charge, their illogical arguments were set aside and the administration were convinced. When the matter was taken up with the Chairman, Railway Board by the Lt Governor of A&N Islands, a similar vague answer was received from M/S RITES without clarifying as to how the soil classification was done. Later while examining old records, it was found that RITES had indeed carried out soil investigation all along the length of the extended portion of the runway by drilling 12 bore holes. Test reports of all the bore holes were available in

APWD records. M/S RITES failed to use this data/information for the purpose of soil classification for the earthwork. The bore holes data clearly showed refusal strata after about two to three meters of depth at different locations. With this document as evidence, the RITES stood completely exposed. For this deficient service they were later debarred from further working in the A&N islands.

3. SCOPE OF WORK

A further examination of the DPR showed that the total earthwork excavation of 37 lakh cum was to be done at three different locations; the first location was along the extended potion of runway, the second location was to lower a hill (known as CARI hill) and the third location was another hilltop where the water treatment plant of Andaman PWD was located. The earth cutting in the last two locations was primarily to remove hill obstructions protruding into the approach funnel. On detailed inspection, it transpired that the CARI hill had nearly 68 staff quarters of Central Agriculture Research Institute (CARI) as also some offices and a garage located on that part of the CARI hill which had to be lowered. Thus, unless those staff quarters were dismantled and relocated elsewhere, lowering of hill could not be done. Similarly, no action had been initiated to construct an alternate water treatment plant before one could consider dismantling the existing one.

4. SELECTION OF PROJECT TEAM

(i) The Chief Engineer was of the view that for a project costing around 50 crores a dedicated project team was necessary. So far only a junior engineer was looking after the airport work. On selection from the Executive Engineers in the Andaman PWD, Shri PK Singh was considered most suitable to lead the team for this project. Similarly, a dynamic and knowledgeable superintending engineer was needed to steer the project for which Shri BN Nagaraja, from CPWD already working in Port Blair was picked up. Suitable Assistant engineers and Junior Engineers were selected with their consultations.

(ii) This team had no experience of handling a runway project. Er. Dogra was of the view that unless the team is made aware of the complexities involved in handling a project of this magnitude, the deliverance might not be up to the mark. Accordingly, a proposal was made to the Lt. Governor and Chief Secretary A&N Islands to allow the entire team to travel by air and visit an ongoing airport project in mainland. With the assistance of Shri CV Nair, the then Executive Director (Engineering), Airports Authority of India (AAI), the entire project team could make a week long visit to a runway under construction in Calicut in Kerala. The visit was an eye opener to the team. They could learn for the first time how a total station equipment is operated and how the levels can be directly transferred in soft form. They saw for the first-time operation of vibratory compactors, tandem rollers, electronic sensor pavers, hot mix plant etc and various other equipment which are so essential for a quality job. They also familiarized themselves with the testing equipment installed at the field laboratory. They observed various tests being done on materials including testing of the core samples, pavement etc. Suffice it to say that when the team returned, it was bubbling with energy and confident to execute the project.

5. ECOLOGY OF ISLAND AND AVAILABILITY OF MATERIALS

It is necessary to elaborate about the ecology of these islands. Every year these islands receive nearly 3100 mm rainfall resulting in high humidity most of the time. There is no known source of sand good for construction in these islands. Local people used sand dredged from sea which is full of salts and coral dust and is thus unfit for any quality construction. The quality of stone aggregates produced by local crusher operators was also not proper with high percentage of flaky aggregates. Other materials like cement and bitumen needed to be shipped from the mainland which is nearly 1200 nautical miles away from Port Blair. In summers, it used to become very hot and sometimes ground temperatures touched 40 degrees centigrade. To handle such heat and high rainfall, the grade of bitumen to be used in the pavement work was modified from 80/100 to 30/40. This was done to ensure that the bitumen did not soften during extreme summer heat and also could handle heavy rainfall. Time period for execution of work decided keeping in mind the maximum dry season available. Daily flight schedules and defence flight operations also had to be kept in mind. Tender conditions were so drafted that the entire work would be carried out during night-time only without disturbing the daily flight operations. It was proposed that the runway will be taken over by the contractor by 5 pm every day and after working through the night, hand over runway by 4 am the next day after providing suitable ramps and mandatory painting etc. Such an arrangement was accepted by the Indian Navy after initial resistance.

6. CALL OF TENDERS FOR PAVEMENT AND RELATED WORKS

(i) As mentioned earlier, the work of Earth work had already been awarded. Similarly, the work of RCC culvert was also awarded. The other three items of work as per the sanctioned estimate were pavement works, boundary wall and the drainage work. It was decided to call tenders after combining all these three sub-heads of work so that the entire project could be covered. As per the DPR, the existing pavement was to be strengthened to PCN 34 to safely handle Airbus 320-200 type aircrafts by providing 200 mm bituminous macadam (BM) in 3 layers, and 50 mm layer each of semi dense asphaltic concrete (SDAC) and dense asphaltic concrete (DAC) respectively (total thickness 300 mm) over the existing pavement. In the extended portion of the runway, over a subgrade of CBR 3%, soil stabilisation with lime and cement was to be done first followed by wet mix macadam (WMM) in 4 layers and then one layer each of BM, SDAC and DAC with total pavement thickness of 1000 mm. Accordingly, the NIT was prepared and approved for call of tenders. The tenders were received in two covers; first cover was to check the eligibility criteria of the bidders which was laid down as per the CPWD manual. The financial bid of only eligible bidders was to be opened.

(ii) During pre-bid meeting, some contractors argued that the time allowed for the work being very short, high rates were likely to be received unless a longer time period is provided for completion of the work. After consultations, bidders were told to quote their rates based on the time period as mentioned in the tender documents only. However, bidders wanting a longer time period were given an option to offer a rebate on their quoted rates. Ultimately no bidder took up the alternate option of longer time period.

(iii) While evaluating the eligibility criteria, there was disagreement in the interpretation of this criteria between the project team and the planning unit team of CEs office lead by Shri AM Kudari SE (planning). Since there was impasse, ultimately, with the consent of the Secretary (Finance), it was decided that only those agencies which become eligible as per both the interpretations of evaluation criteria would be taken upfor opening of their financial bids. This boiled down to only 2 bidders. The lowest bidder was a joint venture between BSC and C&C who were awarded the work in February 1999, after approval of the Works Advisory Board. With the consent of the Administration, a stone quarry was allotted to the contractor to put up his own crushing plant for supply of suitable sized and good quality stone aggregates and crushed sand for the work.

7. RCC CULVERT

(i) There was a large storm water drain at the end of the existing runway. In the sanctioned estimate there was a provision of Rs 0.88 Crore for providing a RCC box culvert at the same location. The width of the culvert was only 2.75 meters with one went way. After careful examination of the area, it was felt that this would not be enough to cater to the discharge of the drain especially during heavy rains. In addition, there were two other issues not considered by the consultant. Discharge was into sea creek and not to any free outflow as is normally the case in the mainland. Second, the tide levels dictate the depth of effective discharge as against physical depth of vent way. Outflow discharge levels were thus not under anyone's control and the discharge capacities could be increased only by increasing the width of flow. With such necessary hydrological designs, additional vent ways were added. In addition, three large water mains of Defence and PWD were passing through the extended portion of the runway and these water mains had to be rerouted through vent way of the culvert.

(ii) While sanctioning the project, the Ministry of Civil Aviation had allowed an upward gradient 1 in 150 in the extended portion of runway to reduce the extent of cutting in the hill sections. On further examination, it was visualized that in the extended runway, the touchdown point may be close to the RCC culvert or even on top of it. The RCC culvert of course was designed to take the impact load of these aircrafts. However later heavy aircrafts were expected to land on this portion of the extended runway. As an engineer, Shri Dogra felt that there should be sufficient cushion on top of the culvert slab to comfortably handle the impact load of any such type of aircraft. It was therefore decided to shift the location of the RCC culvert by 150 meters which provided an additional cushion of 1 meter depth. By implication this also helped in raising the bed of the culvert facilitating city drainage even during high tides. Further, this shifted location allowed engineers to work in dry conditions. RCC brackets on either side in each vent way were introduced to support and carry across various services like water pipelines, electric and telephone cables etc. An open drain properly pitched was connected to the vent ways for uninterrupted flow of water on both sides. This was as one of the important technical contributions to this project by the team.

8.   REHABILITATION WORKS

In the extended portion of the runway there were 212 dwelling units and some shops which had to be relocated. A total of 374 families were displaced. Total land requirement was 48.85 hectares out of which private land requiring acquisition was only 25.32 hectares, balance was Govt land. There was a provision of Rs 5.76 Cr for compensation to be paid to the evictees. For relocation of these families the administration had identified a separate area known as 'Atom Pahad' having an area of around 12.50 hectare. At this location, the PWD took up development works like construction of black topped roads including construction of retaining walls and culverts etc, laying water supply lines, laying barbed wire fencing separating this area from the nearby defence land etc. In addition, shops were constructed at two locations to rehabilitate the affected people. This was necessary as rehab becomes politically very sensitive. Every care was taken to meet the needs of the families who were affected by this project work. The Secretary Civil Aviation specifically inspected this resettlement site to see that rehabilitation work was not neglected by the project team.

9.   DEVELOPMENTS DURING PROGRESS OF WORK

(i) Since this was the largest and most important work of the A&N administration, it was decided by the administration to constitute a high-powered committee (HPC) chaired by the Lt Governor himself to monitor this project. This committee had all the powers to approve any work including sorting out interdepartmental issues especially regarding acquisition of land and allotment of quarry etc. The HPC included the Chief Secretary, Principal Secretaries of Finance, Civil Aviation, and Shipping and all HODs. From Indian Navy side, it was the Fortress Commander himself along with his staff. This committee was a boon to the project as most inter-departmental issues were resolved expeditiously.

(ii) The new contractor for pavement work quickly mobilized resources. He was allotted a quarry with permission to carry out blasting etc. He separately set up a Crushing Plant near work site to manufacture sand and stone aggregates. The contractor also managed to get special permission from the Home Ministry, GOI to bring in a ship load of 30/40 grade bitumen from Iran specially manufactured as per required specifications. He brought the entire stock of bitumen to the site in one go. The contractor also brought all the machinery and equipment needed for execution of the work. He also set up a field laboratory to test various materials etc. To arrive at the optimum job-mix formula, assistance of Shri S P S Bakshi, a senior engineer from the Airports Authority of India (AAI) was taken. He personally tested the job mix formula using different percentages of bitumen to arrive at the optimum mix for achieving maximum density. For resurfacing work, the contractor was asked to take levels of the existing runway surface. To dismay of PWD, large undulations surfaced in the existing runway which required laying of a pavement correction course which was not part of the BOQ. Since final surface finish was the responsibility of this contractor, this extra item had to be executed. The Navy officials co-operated fully and every day by evening the runway was taken over from them and handed over next day early morning to the satisfaction of the Navy.

(iii) Regular HPC meetings were held to monitor the progress of work including resolving various inter-departmental issues. During August 1999, the then Secretary Civil Aviation, GOI, inspected this project along with DGCA and other officers of AAI like the Executive Director (Engineering), GM (Planning) etc. The team inspected each and every part of the project and had meetings with the LG and HPC members. During the HPC meeting, the Fortress Commander of the Indian Navy demanded that considering the strategic location of Port Blair there is need to further strengthen the runway so that Airbus AB 300 series of aircraft can land. This meant strengthening the runway to PCN 60. After detailed deliberations and discussions, this point of view was accepted by the Administration. Based on this decision, the AAI forwarded a revised pavement design to cater for this additional load. As per this design, 280 mm of additional layers (4 layers of BM) was to be laid on existing runway and 375mm of additional layers was to be laid on the extended portion of the runway as well as the turning pad at 22 end. This led to further increase in Revised Estimated Cost. AAI further directed that the Port Blair runway is planned to be equipped with Instrument Landing System (ILS). International standards for such airports require an approach not steeper than 1 in 50 for runway code 4. Therefore, obstructions penetrating the approach funnel are required to be removed to make the full 3300 m runway operational. It was further directed that the runway lighting has to be provided as per design even though there was no provision for the same in the original estimate. AAI agreed to supply all such details/specifications.

10. REVISED ESTIMATE

(i) As the work progressed, it was realized that several items of work were not included in the sanctioned estimate. Unless these items of work were taken up, the project could not be completed. In addition, there were items which got included due to directions from HPC and the Ministry of Civil Aviation. These additional items are listed below: (a) Reconstruction of the residential quarters on CARI hill and construction of a new water treatment plant in lieu of the water tank dismantled; (b) Diversion of existing roads disrupted due to extension of the runway; (c) Construction of new roads to establish links with the existing roads at three different locations; (d) Compensation for acquisition of land acquired for construction of new roads; (e) Additional cost due to strengthening of the pavement to land Airbus AB300 series of aircrafts, pavement ccorrection course and runway lighting etc.

(ii) The revised estimate worked out to Rs 118 Crores which was submitted for sanction to the Ministry of Civil Aviation as per the special format to fulfil various requirements of Public Investment Board (PIB). The Revised Estimate was very well prepared and included justification of each and every additional work. During the first PIB meeting some members unhesitatingly stated that it is one of the best and very well-prepared documents they had seen in recent past. Meanwhile the work was progressing, and the expenditure was nearing the original sanctioned amount. The audit objected to such large deviations stating that this deviation is a major lapse on the part of the Department. Not satisfied with departments replies, the matter was raised by the audit up to the Public Accounts Committee. PAC concerned with such a large increase in the estimate fixed a hearing at Port Blair when some Members of Parliament came over to Port Blair to hear the views of Administration. Shri Dogra CE was asked to defend the case. He was repeatedly questioned on various aspects of the deviation in the work leading to such increase in the estimated cost. During this hearing, all the facts were placed before the PAC with supporting documents including the details of incorrect original estimate. After a marathon session of questioning, finally the PAC was convinced of the logical reasons for such deviations/variations and thus all the objections of the PAC were dropped to the relief of the A&N Administration. This experience and the preparation of such a large revised estimate on PIB format were the highlights in documentation of this project.

(iii) The revised sanction was taking time. The department had almost spent the entire sanctioned amount. The finance department refused to further release funds unless the revised sanction is received or at least a go-ahead signal is received from the Ministry. Apparently, the word spread that the airport work was stopped or suspended for want of revised sanction/funds. A PIL on this matter was filed by a few local residents in the circuit bench of the Calcutta High Court at Port Blair. The department filed details before the court that the A&N

administration has submitted the revised estimate for revised sanction to the GOI and the matter rested with the Civil Aviation Ministry of GOI. In the next court hearing, the Ministry was asked to make their submissions. Finding no fault with the department the Civil Aviation Ministry gave a go ahead signal to continue with the work pending receipt of formal sanction. With this clearance, the work could further be proceeded with full gusto and later completed. A visit to the islands by Shri PV Jayakrishnan, Secretary, Civil Aviation, GOI was a great morale booster to the project team as the work was appreciated by him. He gave detailed directions to his officers in all aspects of work and assured the administration that his Ministry would provide timely support. As mentioned earlier also, Shri CV Nair, the then ED (Engineering) AAI and Shri SPS Bakshi GM AAI also made very vital technical contributions and gave valuable tips/ guidance in completing the project. Without their assistance it would have been difficult to do quality work and complete this project. From the Indian Navy, Admiral Harinder Singh, the Fortress Commander and later Admiral Raman Puri, Fortress Commander attended all HPC meetings and gave exemplary support in day to day working and handing/taking over of the runway. Because of their unstinted support, the entire defence establishment of these islands provided assistance whenever asked for.

11. INSPECTION BY CTE

After all such issues were resolved the work was progressing well. But the Hon'ble MP of the Islands kept on bitterly complaining to the LG against the project stating that funds were being siphoned off and that there was rampant corruption and that very poor quality of work was going on in these islands. Apparently, he had not been properly briefed. The Chief Engineer was his main target. With this incessant complaining, the Chief Engineer offered to quit as in any case his deputation period was nearly over. But the administration stood by the Chief Engineer and asked him to continue working without any fear or favour. Later fed up with this incessant complaining, the LG, taking the CS and Chief Engineer into confidence, wrote to the Central Vigilance Commission (CVC) to get the project inspected technically to satisfy the doubts of the MP. In view of this request, a team of engineers of Chief Technical Examiner of CVC (CTE) inspected the project during June / July 2000. The team collected core samples of the pavement at different locations to verify the thickness of pavement layers as well as the bitumen used. They also collected samples of all the materials, examined all the work records right from the inception till award of work. The report of CTE was received much after Shri Dogra was repatriated to his parent department, i.e., CPWD in August 2000. Later the A&N administration invited Shri Dogra to defend the revised estimate before the PIB team which was visiting the islands. This visit was also utilized to prepare replies to most of the CTE paras. CTE could not locate any deficiency in either the thickness in pavement layers or in the quantity of bitumen used in the work, or in quality of any material. Having found no deficiency, only general observations were made like; why the three subheads of work were combined to call a single tender. They also made some factually incorrect conclusions that the justification of the tender prepared by the division, the circle office and zonal office were same which was factually not correct. Each office had prepared their own justification and finally the one prepared by the planning unit attached with the Chief Engineer's office was only accepted and this was the lowest. Suffice it to say over a period of time all the CTE paras were dropped. This itself speaks volumes of the efforts and the hard work put in by the entire Andamans PWD team of engineers who worked day and night on this project. The vigil maintained by the team paid off as despite their best efforts, the CTE team could not point out any deficiency in the work. This APWD team worked day and night and therefore deserve to be recognized and honored. This includes S/Shri AK Bardhan, D Balaji, Jose P John, Gautam Chakravarty, BK Singh and L Gandhi.

12. COMPLETION OF PROJECT

Despite a number of constraints/problems related to preparation of the project and its execution, the APWD team managed to deliver a high-quality runway work for the benefit of the islanders. The benefit of this strengthened runway was specifically realized in December 2004 when a huge tsunami hit these islands among other parts of the country and the world. Several buildings collapsed or were severally damaged at Port Blair including some cracks appeared in the Airport terminal building. Thankfully there was no damage to the runway. Bringing relief material to Port Blair became utmost important for GOI. The extended and strengthened runway became an important lifeline for relief operations when the Indian Airforce landed one of their largest aircraft 'Gajraj' without any difficulty. More than 40 flights landed on a single day carrying relief materials. Thus, there could not have been a better certification for the quality and utility of this runway in such diverse situations.

Shri Dogra have acknowledged that this project was possible with full support by the local Administration. Further patronage and guidance were given by Shri Ishweri Prashad Gupta, Hon Lt Governor. Strong support in administrative and financial matters were provided by Shri R Narayanaswami, Chief Secretary. Shri Xitij Kumar Mahata, Dy Commissioner of Andamans Island gave prompt backup to ensure quick land acquision, quarry allotment, rehabilitation matters etc. Shri Pradeep Singh, IAS, Secretary PWD, was instrumental in finalizing the revised project estimate and gratitude is due to him.

LESSONS LEARNT

(i) **Coordination for project and role of stakeholders:** The Andaman PWD was the implementing agency which had to carry out the works according to the technical parameters laid down by the Airports Authority of India under the Min of Civil Aviation, with physical control of the Indian Navy and under the administrative coordination of the Lt. Governor, A&N Islands. All this essentially required tremendous amount of coordination among different wings of the government. The support and patronage by Administration under the guidance of Lt. Governor, through High Powered Committee, helped in ensuring faster decision making, resolving issues and completion of the project.

(ii) **Significance of Project Preparation:** Project preparation is an important activity. The work done by consultants should not be taken for granted. It is necessary to examine their work and take corrective action. The experience of Shri Dogra did help in taking up issues & resolving the same.

(iii) **Selection of Team and team spirit for success of project:** Forming a team and providing hands-on training to its members, ensured smooth execution of the project. It also gave them a sense of commitment, bonding and confidence.

(iv) **Project uniqueness and related challenges:** Challenges and large scale changes at the execution stage usually tend to create administrative and contractual problems and need to be avoided as far as possible. However, projects like strengthening and extension of runway do not come every day. Proposed upgradation to handle AB 300 series of aircrafts and providing ILS facility was a forward looking action which enhanced the utility of the airport.

It is a unique case where in the revised Administrative Approval and Expenditure Sanction was expedited by Ministry of Civil Aviation under public pressure through a PIL filed in Calcutta High Court.

(v) **Right outlook:** Examination of works by the CTE or any such body is not a criticism but need to be taken as a certification of the systematic and quality work done by the team.

UNDERPASS AT MADHUBAN CHOWK ON OUTER RING ROAD, DELHI

1. AS ONE OF THE IMPROVEMENT MEASURES

As one of the improvement measures, it was decided to start with provision of grade separators at every junction of Outer Ring Road, so that slow moving traffic can move with speed without any interruption. In North West Delhi, Rohini, the biggest urban extension is beyond Outer Ring Road. One of the important intersection for traffic movement to Rohini with Lala Jagat Narain Marg (Road No 41) is known as Madhuban Chowk. At this intersection there was heavy traffic and travel time on both roads was excessive. It was decided to consider providing under/over bridge to ease the traffic movement. At this location, across Outer Ring Road, overhead Metro line was already under construction. Providing over bridge above the Metro line was not feasible and possible as it would have been very high and approaches on both sides would have been very long. Therefore, only feasible solution was to provide underpass along Lala Jagat Narain Marg. Accordingly, preliminary investigations were initiated.

2. SUBSOIL INVESTIGATIONS

During subsoil investigation, it was found that for considerable depth, there was alluvial soil and underground water table was about 2 meters below the ground level. According to geometric standards, the Underpass was required which would be about 7 meters deep. Thus, base slab was to be subjected to about 5m uplift due to sub soil water pressure. In order to counteract this uplift pressure, it was decided to provide permanent pre-stressed soil anchors. After field survey, the geometrics of underpass were decided. The underpass, having a length of 555 m, was an underground tank like structure, planned across Outer Ring Road. The Underpass was designed to cater for two-way traffic with 9 m wide dual carriageway, each side having three lanes. The Underpass was to be covered with 1200 mm thick pre stressed voided slab with a clearance of 5.3 meters for the traffic movement. For this 60 m long underpass camber of 2.5% was provided for drainage of rain water. The longitudinal slope was designed for mixed traffic on Outer Ring Road at 1 in 30. One arm of underpass was in curve and super elevation of 4 % was provided and in the straight portion camber of 2.5% was provided for drainage. Safety crash barriers were provided on both sides of each carriage way. Suitable sump and pumping arrangement with power backup provided to drain out the rain water by pumping.

3. STRUCTURE AND CONSTRUCTION ASPECTS OF THE UNDERPASS

The construction of the Underpass was taken up from top to bottom (top down method). The side retaining walls were designed and constructed as Diaphragm wall in panels of 5 m width having thickness of 800 mm at deep portions and 600 mm at shallow portions. As a part of top to bottom construction, first of all 12 panels, i.e., 6 on either side, 800 mm thick diaphragm wall panels of closed portion were cast followed by the casting of slab at the required level on the virgin soil only. While placing the shuttering plates for the casting of slab, special attention was paid to put them in a pre-designed pattern with well-defined joints so that the ceiling of covered portion after the construction of Underpass is aesthetically pleasing. After casting of all diaphragm wall panels on both sides of the Underpass, the excavation was carried out to remove the earth. Since the water table in the Underpass area was too high, dewatering wells were provided at close intervals and pumping carried out round the clock to keep the water table low and ease out the construction activities like excavation, waterproofing, laying of base slab and installation of anchors including pre-stressing and grouting etc. After successful installation of the anchors, remaining cosmetic works like wearing course, cladding to diaphragm wall with decorative finish, crash barriers, planters etc. were executed.

4. TECHNOLOGIES FOR CONSTRUCTION

Technologies for construction adopted were diaphragm walls for construction as retaining wall, pre-stressed voided slab in covered portion spanning 23.5 m between the diaphragm walls, use of mechanical couplers for joining reinforcement of base slab and diaphragm wall, poly-fibre reinforced concrete (PFRC) wearing course with the vacuum dewatering system for reducing water- cement ratio and gaining the higher strength fast. The active soil anchors were provided for the first time in India, to counter uplift pressure on the base slab on account of high water table. Conventional solutions like providing tension piles or adding weight to base slab by heavier material like iron ore were also considered. However, use of soil anchors was considered preferable.

5. SELECTING TYPE OF SOIL ANCHOR

Depending upon type of soil strata different types of ground anchors are provided. Rock Anchors are provided if there are rock strata below. If only soil strata existed for considerable depth, soil anchors are provided, as was done at this location. On account of high water table, considerable elastic stretch was likely to be caused to strands. Relatively large movement of anchor head was required to mobilize the full load carrying capacity of anchor. To reduce such movements the anchors were prestressed. Totally 992 soil anchors were provided in the area where slab was subjected to uplift pressure. Each Anchor was designed for 40 T capacity. The location of anchors was decided depending on uplift pressure expected. In shallow areas spacing was at 3.5 m whereas in deeper area spacing was reduced to 2.5 m.

6. ASSESSMENT OF CAPACITY OF SOIL ANCHORS

Assessments of design capacity of Soil Anchors was necessary for a safe design. To assess capacity of Soil Anchors and decide parameters, initial field tests were conducted. The anchor was to be connected to bottom of the underpass which was about 7 meters below the ground. The required free length of anchor below the underpass was about 5 meters. Thus, for testing at ground, free length was taken as 10 m. The free length was increased to 17 meters to account for depth of underpass (7 m), and fixed length of anchor was taken as 10 m. These anchors were tested in accordance with FIP recommendation and 40 T design load was taken. The Performance Test, Proof Test and Creep tests were conducted as per standing guidelines for pre – stressing system. The anchors could not pass through any of the trial tests. It was observed that the movement of the anchor at maximum load was required to be reduced and to restrict this movement, it was necessary to increase the friction between the anchor and soil interface. In the test anchors, the anchor duct was plain in the upper 10m i.e., unbonded length and corrugated in the lower 10m i.e., bonded length. In order to increase the frictional resistance, the anchor duct was made totally corrugated and were tested again keeping (a) Bonded length of anchor to 10 m, (b) Unbonded length also to 10 m, (c) Diameter of bore hole as 150 mm, (d) Maximum Ultimate bond stress between cement and grout and surrounding soil as 2.5kg/cm^2, (e) Factor of Safety for bond strength between grout and surrounding soil as 3 and (f) Factor of safety for tensile stress as 2. These anchors were tested in accordance with the FIP recommendations and were found to meet the performance criteria.

7. LOADING CONSIDERATIONS

Since the base slab was cast after installation of diaphragm wall and excavation of earth, the earth pressure on diaphragm wall as well as the self-weight of diaphragm wall did not get transferred to the base slab. The base slab spans transversely between the diaphragm walls and rested on soil which provides flexible support. Since the base slab was cast over the ground, the self-weight was assumed to be directly transferred to the soil sub-strata. The reaction on top of diaphragm wall due to superimposed dead load and live load on top of cover slab also induce moments in the base slab. The soil anchors were spaced along the traffic direction at 2.5 m centre to centre. Eight such anchors were provided along transverse direction. Analysis carried out with all the anchors present as well as for eight different cases considering one anchor out of service, due to its failure. For estimation of hydrostatic pressure, highest water level was considered at 1.0 m below the ground level. It was found that the hydrostatic pressure acting over base slab decreases the base pressure due to soil anchor force imposed on it, as they act opposite to each other. Part of soil anchor force got counteracted by the hydrostatic pressure and balance carried by soil springs as base pressure. When the hydrostatic pressure was absent during the condition of lowering of ground water table, higher base pressure was required to balance the soil anchor force. In fact, the presence of hydrostatic pressure only altered the distribution of pore pressure and base pressure without any significant change in the base slab design moments.

8. FABRICATION OF ANCHORS

Basically, each unit of the soil anchor provided in this underpass was 20 m long, consists of 5 nos., 12.3 mm low relaxation high tensile steel strands encased in 80 mm diameter High density poly ethylene (HDPE), 4 mm thick, doubly corrugated tube. Out of 20 m length, lower 10 m length was made as bonded length and upper 10 m as un-bonded length to allow the elongation in strands, when pre-stressed after the installation. Epoxy paint was applied on complete length of the strands. In the bonded length, it was followed by the sprinkling of fine silica sand to ensure a good bond, while in the upper 10 m length, i.e., un-bonded length, all the individual strands were encased in plain HDPE sleeves after application of grease. This made the movement of anchor free and created a non-corrosive environment around the strands. The junction of bonded and un-bonded lengths of each strand was sealed to restrict the penetration of cement grout into the un-bonded length during the grouting operation and to prevent the grease from oozing out. Complete assembly of the five strands properly spaced by means of poly vinyl chloride (PVC) spacers and supported on centralizers were encased in 80 mm diameter corrugated HDPE tube, making one complete unit of the soil anchor. The lower end of the anchor was closed with a PVC cap and sealed with adhesive. Spacers provided at 3m centre to centre were placed to separate each of the five strands uniformly, forming the inherent part of the anchor after the anchor was installed and grouted. Use of centralizers ensured a grout cover of minimum 10 mm over the anchor bond length.

9. ENHANCING DURABILITY

It was essential to protect the strands from getting corroded to have a longer life of the anchor and so the structure. Due to presence of high-water table in the area, the anchors were surrounded in a fluid environment. Therefore, special treatment was done for the corrosion protection by providing multiple protections to strands in bonded as well as un-bonded length. In the bonded length, the strands were applied Epoxy paint and the inner as well as outer sides of the HDPE ducts were cement grouted to prevent the entry of water into the strands. Similarly in the un-bonded length, the strands were Epoxy painted and wrapped in grease, encased in plain HDPE sleeves and again the inner and outer sides of the HDPE ducts were cement grouted. While applying the grease to the stands in the un-bonded length, the strands were uncoiled a little to ensure that the interstices between the wires of seven-wire strands were completely full of corrosion inhibiting grease. While the anchor was fabricated, one 20 mm diameter, plain HDPE sleeve was left inside the anchor and sufficiently projected out of the anchor length, so that it could be used for pushing the grout during installation of anchor. In addition to this, one more similar sleeve was placed outside the anchor for external grouting.

10. INSTALLATION OF ANCHORS

Reinforced Cement Concrete (RCC) Base slab 800 mm thick in the deep portion and 600 mm thick in shallow portion, was cast prior to the installation of the anchors. While the base slab was cast, a through hole of 300 mm diameter was created at the anchor locations and after the concrete was set and gained sufficient strength, 150 mm diameter bore in the soil beneath the base slab was drilled using rotary drilling machine. For the proper functioning of anchors, it was essential to maintain the friction between the anchors and surrounding sub-soil strata and therefore bentonite was not permitted. Initially, it was experienced during the installation of test anchors that the bore hole after drilling was getting collapsed as no stabilization system was adopted. Hence, while drafting the specifications for working anchors, it was made mandatory to use full length casing till the prefabricated anchor was housed and then the casing was withdrawn while grouting. This was made possible with the use of Casagrande make C-6 Rig, in which the cutting tool was fixed on the casing itself. This machine performed the drilling and casing simultaneously and thus leaves no chance for the collapse of the bore. As the bore work progressed casing was added in 2.5 m lengths by coupling with the previously inserted casing in bore hole. Once the borehole is completely drilled to the require depth, it was flushed with water and thereafter the anchor was lowered into the borehole manually followed by the grouting of the bore hole. While the grouting was in operation, the casing was withdrawn in parts in reverse manner.

11. GROUTING

Grouting of anchor and borehole was adopted as one of the corrosion protection measures to protect the strands. The anchors were grouted internally as well as externally with cement grout having a water-cement ratio of 0.4. In order to control the shrinkage, suitable admixtures were added in the grout. It was ensured that the grout produced was free of lumps and uniformly dispersed in a colloidal state and also pumped continuously. The grout was injected from the lowest point and progressed upwards. The grout pressure was controlled to prevent excessive heave or fracturing. Grouting of the soil anchors was done in two stages. In the first stage only the bonded length of the strands was grouted so that un-bonded length was free for taking the movement of the anchor during the stressing operation. The space inside the HDPE sheathing was grouted up to the top of bonded length and the annular space between sheathing and surrounding soil was grouted up to 1m below the bottom of the base slab. After stressing and testing the anchors, the un-bonded length inside the anchor and remaining length outside the anchor were pressure grouted. At the end of the grouting, the grout tube was finally filled with grout to avoid the presence of any air inside the anchor. It was observed that in 21 days, the grout gained sufficient strength and also necessary friction developed between grout and soil interface for transfer of stress to the ground.

12. AFTER GROUTING

The anchor was pre-stressed after a minimum of 21 days. The bearing plate and anchor head were placed in the pocket created in the base slab during its casting. The bearing plates were so placed such that the axis of the anchor was normal to the bearing plate and passed through its centre. After the anchors were installed at site, all of them were tested to ensure their successful installation. Performance test was conducted on 5% of the anchors while the proof test was performed on remaining 95% of the anchors. After the completion of the pre-stressing and the pressure grouting, injection-grouting was carried out to effectively seal any loose pockets and eliminate the possibility of oozing of water. All anchorages were epoxy painted and covered with grout after stressing. The recess was then concreted with non-shrink grout. The base slab was overlaid by 75 mm thick poly-fiber reinforced concrete wearing course for a smooth riding surface and as a protection layer to main base slab as well as soil anchors.

13. CONCLUSION

Underground structures have distinct advantage of maintaining the overall environment on the ground without causing any unpleasant obstruction to the skyline. Creating an underground structure may require a bigger deal with technical inputs, but the solutions are available, though may be slightly costlier compared to the structures above the ground. But the savings extended to the environment always overweigh the additional cost caused due to the Underground structure. At times, when the situation like Madhuban Chowk necessitates the construction of an underground structure, it is possible to design and implement such structures, especially in urban environment with the use of latest technologies available worldwide. Though a conservative approach may result in implementation of conventional technologies, but the latest technologies always provide more efficient and cost-effective solutions. Use of permanent prestressed soil anchors in the construction of Underpass at Madhuban Chowk proved to be a cost effective as well as time effective solution to counter the high uplift water pressure exerted on the base slab.

LESSONS LEARNT

(i) **Adoption of new technology and innovation:** By adoption of New Technology the Engineers of Delhi Tourism and Transportation Development Corporation Limited (DTTDC) did outstanding work under leadership of Dr Shishir Bansal. The courageous initiatives are worth appreciation. Even after almost twenty years, the project is serving its purpose. This shows that their initiative opened new avenues in technology. Also appreciate DTTDC for implementing this project with new technology and thus showing full confidence on engineers, who belonged to CPWD Cadre.

While adapting new technology, courage and confidence are necessary. Besides, guidance from an experienced structural designer was necessary and Late Shri T N Subarao was the right person to guide.

(ii) **Inspiration for future Engineers:** This project should be visited by young engineers to understand new technology. Our country can advance only if we accept advance technologies for our projects.

Underpass at Madhuban Chowk on Outer Ring Road, Delhi -
9 m wide dual carriageway underpass along Road 41.
The Outer Ring Road is at grade. Elevated Metro line is also seen

Underpass at Madhuban Chowk on Outer Ring Road, Delhi -
View of the Underpass

Underpass at Madhuban Chowk on Outer Ring Road, Delhi -
Another View of the Underpass

1. BACKGROUND

The "Signature Bridge" a new landmark in Delhi, the capital city of India, is India's first asymmetrical cable-stayed bridge, and connects the main city with East Delhi across the river Yamuna. The Western approach of the Bridge connects the nearby areas like Timarpur, Nehru Vihar, Wazirabad, Aruna Nagar, Mukherjee Nagar etc. and Eastern approach connects to Mangal Pandey Marg in East Delhi. In a meeting held during July 1998 at Raj Niwas, under chairmanship of Hon'ble Lt. Governor of Delhi, it was decided that the feasibility of constructing the new bridge on river Yamuna, as a barrage cum bridge with higher pondage level should be explored. On examination, it was noted that composite barrage-cum-bridge would result in higher pondage, but it would submerge fresh areas having inter-state implications. Besides, shifting and rehabilitation of affected people would add to the cost and delay the project. In Feb. 2000 it was decided to construct a new bridge at the downstream of the existing Wazirabad bridge. Meanwhile, National Highway Authority of India (NHAI) had also proposed a single four-lane flyover over National Highway-l in its vicinity. In December 2001, Hon'ble Minister PWD expressed his desire to integrate the proposal of NHAI with the proposal of Government and submit to the Technical Committee of DDA. Decision also was taken by Govt. of Delhi to de-link the construction of the barrage from the construction of the new bridge. Accordingly PWD decided to explore the new alignment for the proposed bridge. Delhi Tourism & Transportation Development Corporation Limited (DTTDC) submitted their proposal in November 2003 to construct this bridge as a Deposit work. The offer was accepted by Govt. of Delhi. A Memorandum of Understanding was signed between the Government of Delhi and DTTDC on August 27, 2004. DTTDC appointed M/S STUP Consultants Pvt. Ltd. as consultant and got the feasibility studies conducted. Based on the feasibility studies, a proposal was framed and submitted for approval of Technical Committee of DDA. For making it a landmark structure as also a tourist destination a proposal given by M/s Ratan J. Batliboi, Urban Planners and Architects Schlaiachi Bergemann & partner M/s Construma Consultancy Pvt. Ltd. was approved. It was decided that proposed bridges would be Cable Stayed Type with Central Pylon having 150 meter height (double the height of Qutab Minar). Viewing galleries, restaurant and leisure activities were also conceptualized. For visitors, to have panoramic view, lifts upto the top were proposed. Pattern chosen was symbolic to Indian Culture and reflected modern India. The challenge was not only to conceive a bridge that is iconic, but also having a form following the flow of forces thereby, being in harmony with structural engineering concepts at the same time symbolizing local culture and tradition.

2. EVOLUTION OF DESIGN AND APPROVAL

During the phase of evolution of design, many options were considered with the fundamental principles enumerated above. The height of the pylon was decided as almost twice the height of Delhi's another heritage structure Qutub Minar. Graphics on the bridge structure, particularly on the Pylon was perhaps featured first time in the world. The pattern chosen was not only to symbolise Indian culture but also to reflect modern and progressive India. After having evaluated various options, peacock's feather was agreed as an appropriate emblem on the Pylon. Initially a Preliminary Estimate for the project was sanctioned by Delhi Government in March 2006 for an amount of Rs.459.00 crore with a normal balanced cable stayed bridge. Due to change in scope of work, a recast Preliminary estimate of Rs. 993.11 Crore submitted to Delhi Government on in May 2007. This Estimate was not approved immediately by Delhi Government as cost was considered very high. Thereafter, Delhi Government decided to request Shri E. Sreedharan, the then MD, DMRC for his expert opinion. Shri. Sreedharan, after examination the estimate, recommended that cost estimate as proposal was acceptable. He explained that this a very unusual bridge meant to affix signature status to Delhi, an important tourist destination of the country. The bridge was therefore, different from the usual symmetrical designed cable stayed bridges. In February 2010 Delhi Government sanctioned an Estimate amounting to Rs. 1131.00 Cr. with bridge component as Rs. 631.81 Cr. Thereafter, the work on the project started. A substantial portion of site could be made available only after tree cutting permission from Ministry of Environment & Forest, Government of India, in February 2012. It took nearly 5 years to obtain this permission.

3. WESTERN AND EASTERN APPROACHES

The bridge is connected to Ring Road on Western side and Mangal Pandey Marg on Eastern side. On Western side, grade separators comprised of flyovers, loops and ramps which were constructed to ensure signal free traffic movement at the proposed intersection of bridge with Road No.45 and existing intersections at Timarpur, Nehru Vihar and Wazirabad. Road widening, including construction of footpaths and cycle tracks was also part of work on Western approach. Works on Eastern approach Road include construction of Embankment of about 2.0 Km long, river training work, river protection works, widening of existing roads, construction of new road, footpath, cycle track, storm water drain etc. In addition, on the Eastern side, six lane flyovers was constructed at Khajuri Khas intersection with rotary at ground level to ensure signal free movement upto Mangal Pandey Marg. The pre-cast segmental construction technology was adopted for all viaducts in approaches. Due to constraints at some locations, pre-cast construction was not feasible and therefore in situ construction was done. Voided slabs and Solid slab were provided in loop of superstructure. Open foundations resting on rock as also bearing piles, socketed in rock were provided. Self-Compacting concrete of M60 & M65 grade was used for piers and diaphragm wall. The entire structure was integrated with piers, thus avoiding bearings and expansion joints. This arrangement improved the riding quality and made the structure maintenance free. Open foundations / Pile foundations / Well foundations were provided as per design requirement based on sub soil strata. The ramp portion of the grade separators was provided with Reinforced Earth Walls. The work of different components was executed in different phases and opened to traffic as and when individual component was ready. The main flyover on Ring Road on Western side of bridge was opened to traffic in August 2012. Flyover at Khajuri Khas on Eastern side was opened to traffic in March 2014, and on Western side approach opened in April 2015. Balance work though completed earlier but opened to traffic along with bridge in November 2018.

4. PROCUREMENT, FABRICATION AND TRANSPORTATION OF STEEL DECK AND PYLON

According to structural design, steel plates of 20 mm to 250 mm thickness of High-grade steel i.e., S-355 & S-460, were to be provided. Decision was to be taken, from which country, these plates were to be procured and fabricated. Details of Steel plates manufactured in our country were ascertained. It was noted that plates upto 63 mm thickness are manufactured in India. Therefore, procurement from foreign county was necessary. For fabrication three top workshops of the Country were considered. These were (i) B.B.J. workshop Kolkata, (ii) L & T workshop Chennai, and (iii) Mazgaon Deck workshop Mumbai. Out these three workshops, only B.B.J. workshop at Kolkata submitted their proposal. This workshop was visited and work programme as also manner of fabrication discussed with authorities. This workshop was not found suitable.

Thus for fabrication workshops of other countries were to be considered. Another important requirement from logistics was that as far as possible the procurement and fabrication should done form the same country. After collecting information, workshops of M/s Arceler Mittal at Netherland, M/s Jessop, M/s ZTSS at China and M/s Willian hare at Abu Dhabi were visited and shortlisted. Quotations were called and six proposals were received. Out of these M/s ZTSS from China were selected for fabrication. Besides, for procurement of materials, Chinese suppliers were selected for convenience of working. Besides, M/s Lloyd's Resister of U.K. were approved as the third party quality assurance agency, for procurement, fabrication and erection of structured work for pylon and deck of main cable stayed bridge.

For fabrication propose, the pylon legs were divided in 10 different segments each. The Main body of pylon, was divided in five different segments and pylon head considered separately. The five main body segments were further divided in another 23 parts. Each segment was assembled with 10 to 20 sub-components at ground. These segments were transported from China by ship to Kandla Port in Gujarat State (India). From Kandla Port to project site, these were transported by road in heavy duty long truck trailers. It was noted that the bridges on the route were not strong enough to bear the load of truck trailer. For this purpose, either the road was strengthened or had to construct temporary Bridge. It was an extremely difficult task and transportation contractor did in excellent manner.

5. COMPONENTS OF CABLE STAYED BRIDGE

(a) **Foundations of Bridge** - For cable stayed bridge foundations of the following types were provided.

   (i) 6 nos. of open foundations resting on rock strata at a depth of about 20 m. The main Pylon foundation (2 nos.) were 23 m deep with Pier dia. of 5.5 m and lateral spans having foundations (4 nos. each) 7 m deep with Pier dia. of 2 m.

   (ii) 15 nos. of well foundations for the remaining piers, with dia. ranging from 8 to 9 m were provided. 2 Nos. well-pile integrated foundation, 17 m dia specially designed foundation was provided at one location on account of sloping rock and backstay were provided.

     The sub soil strata at site is generally homogenous and comprises mainly of two types of layers, dark grey fine sand and light brown sandy silt up to the rock layer around 20 m below the ground level. The rock met were generally weak to moderately strong quartzite.

(b) **Bearings & Their Layout System** - There were twenty free spherical bearings, two large fixed bearings under pylon legs, two numbers of longitudinally guided spherical bearings at the location of expansion joints to enable reversible movement of 270 mm and rocker bearings under backstay anchor foundation which can transfer the tensile force of 6500 T within the main cable stay bridge system. Typically, the bearings are placed to support longitudinal plate girders on both axes. The bearings are typically pre-set for transverse movement, longitudinal movement due to elastic shortening of the girders and rotational movement due to pre-cambering in the transverse direction. The special attention was given to detail the fixed bearings under the pylon legs at foundation location and rocker bearings where 8 numbers of back stay cables were anchored to foundation.

(c) **Main Cable Stayed Bridge** –

   (i) Main Cable stayed Bridge comprising of a symmetrical inclined Namaste shaped Steel Pylon of 154 m height. Total length of the cable stayed bridge from expansion joint to expansion joint was 575 meters, with main cable-stayed span of 251 meters supported with 15 sets of cables on one side and counterbalanced by 4 sets back stay cables attached at a pendulum bearings. The bridge was made of steel and concrete composite deck. It has dual carriageway of 4 lanes (14 m) each with about 1.2 m. central verge. Space for anchoring cables, maintenance walkways and crash barriers were provided on either side of central verge. The outer to outer width of the bridge was 35.20 m, and approach spans were about 36 m long. Spherical bearings provided on all the piers. Pendulum bearings provided for back stays.

(ii) The Steel Pylon was of 154 m height from top of the bearings consisted of two legs made up of steel boxes which merged into one upper pylon body zone made up of a load bearing member stiffened by internal stiffeners and bracings, where the cables supporting the main span and the back stays were anchored to deck. Each of the pylon legs consists of hollow steel boxes that were roughly 70 meters high. The upper end was the kink diaphragm. It had transition from pylon leg to the pylon body.

(iii) The pylon head was made up of beams and columns in steel structure with a glass cladding. Major portion of the steel for pylon was of grade S355. In highly stressed anchorage zones, S460 grade steel was used. Each leg of Pylon rested on spherical bearings to transmit vertical loads of more than 17,000 T.

(iv) The deck is 32 m wide width for 8 lanes, 4 lanes in each direction. The composite deck consisted of two main girders (I-shaped) in longitudinal direction and cross girders at 4.5 m spacing along the deck. Spans are of 13.5 m long on the cable-supported part, 36 m on the approach spans that supported over concrete columns. Most part of the deck slab was made up of full depth prefabricated concrete elements of varying thickness from 250 to 350 mm, stitched in-situ over steel girder flanges. In highly stressed areas, near pylon base and backstay anchorage, in-situ concrete up to 700mm thick was provided.

(v) The cables made up of bundles of parallel 15.70 mm strands of class 1770 Mpa, protected against corrosion with hot dip galvanization and outer PE-pipes. Depending on the location the number of strands per cable varied from 55 to 123 nos. at the main span and 127 nos. for each backstay.

6. COMPLEXITY OF FABRICATION JOB

Components were fabricated in China, and fabrication work was done by welding and bolting at project site. It was a very difficult and complex task, as pylon was a three dimensioned complex structure having inclination in all the three plans. To study dimensioned weights of the elements to be fabricated and transported a true to scale digital model of the bridge was done in Tekla, a steel detailing software. It incorporates a Building Information Modeling (BIM), enhancing efficiency, accuracy and substantial reduction in wastage of material though proper detailing. Such detailing was done for pylon as also for deck. Detailed planning was done for cutting and joining of plates. Besides, most appropriate welding procedure was adopted, Compression joint were designed to transfer load through bearing contract surfaces. Preassembling was specially designed according to the adjustments in three directions.

7. ERECTION OF PYLON SEGMENTS

(a) Signature Bridge, a cable stayed bridge, was a highly unstable structure during erection. The pylon was divided into segments with weights varying from 40 to 250T. The points between segments were planed bolted joint with machine finish. To erect pylon using a 1250 T capacity crawler crane, and pylon supported with specially designed temporary installation for permanent cables segments weighing between 40 to 250 T were created. Based on the requirement for erection, temporary prop arrangement was designed to support the pylon. To maintain and ensure correct geometry, suitable arrangement was provided. Specially tie down arrangement was installed at the base of pylon to stabilize the initial phase of pylon erection. During construction of independent foundation on both sides, the tie down shades were constructed using prestressing cable. There RCC shades (44 m × 2.275 m) were water proofed and constructed upto ground level. Load was given as per design sequence. After erection of segments and installation of base support, the tie down system was de-tensioned.

(b) Deck girders were supported over temporary trestle and were erected using Goliath gantry. Thereafter precast deck panes were erected. These were stitched by cast-in-situ concrete on top of the main girder. There were 528 precast panels of 4 types, generally having plan dimension of 7.5 m × 4 m with 250 mm thickness. These were cast at casing yard. Stage by stage analysis was done to determines stress distribution in the structure for all intermediate stages. A careful monitoring throughout the construction was done so that target geometry created be achieved. The variation in geometry on account of temperature variation was also accounted for. Survey data was collected and examined for relevant erection stages, based on survey points that were marked during trial assembling.

(c) After installing anchorage boxes, with bearing plates, at pylon top and deck levels, ducts were welded and tubes were hoisted. First 80% stressing was done for all cables and after that temporary stressing was done. As the pylon grew in stature, the access to the working location of bolting, pre tensioning shifting of jacks and other accessories was done by specially designed platforms, cage ladders and passenger hoist equipped tower crane. The connection between pylon and tower crane was engineered in such manner that no unscheduled horizontal force remained on pylon and tower crane. Finally the bridge structure was completed.

8. INFORMATION ABOUT COMPLETED PROJECT

Cost of the Bridge after completion

The total cost of project after completion was Rs. 1518.37 Crore including 5 % as departmental charges. Out of this the cable-stayed bridge has costed Rs. 885 Cr and the approaches Rs 431 Cr. Cost of other major components are Rs. 13 Cr for lifts, Rs. 17 Cr for pylon access system, Rs. 4 Cr for Bridge health monitoring system and Rs. 8 Cr for electrical works.

9. ACKNOWLEDGEMENT

(a) For this project there was direct or indirect participation by experts, technicians as also workshops of 11 countries. these are (i) Australia, (ii) Brazil, (iii) Switzerland , (iv) Canada, (v) Spain , (vi) Luxembourg ,(vii) China, (viii) Italy, (ix) Germany, (x) South Korea and (xi) France. Their contribution is gratefully acknowledged.

(b) Manufacturers of different Components-
    i. Fabrication of steel pylon and deck - ZTSS, China;
    ii. Bridge bearings- Maurer Shone, Germany;
    iii. Expansion joints - Mageba, Switzerland;
    iv. HSFG bolts - TVS Fasteners, Chennai (India;)
    v. Cables - M/s Tensa – Spain;
    vi. HT Railing and steel fabricators - Rajhans - Uttar Pradesh (India);
    vii. Structural steel fabricators - Saifee Steel - Uttar Pradesh (India).

(c) Consultants who have made this Project possible -
    i. M/s STUP Consultants for the Feasibility Studies for the entire Project
    ii. M/s Tandon Consultants Pvt. Ltd. as design consultants for Approaches
    iii. M/s Construma Consultants Pvt. Ltd. as Proof Consultants for Approaches
    iv. JV of M/s Schlaich Bergermann Und Partner (Germany) & M/s Construma Consultants Pvt. Ltd. (India) as design consultants.
    v. JV of M/s Systra SA (France) & M/s Tandon Consultants Pvt. Ltd. as Proof Consultants for Signature Bridge.
    vi. Studio De Miranda for construction stage analysis of Main Bridge
    vii. Lloyd Register Asia, UK for Third Party Quality Assurance of Bridge

(d) Technical Studies -
    i. Hydraulic Model Studies by Central Water and Power Research Station Pune.
    ii. Seismic Studies by IIT Roorkee, UP
    iii. Wind Tunnel Study by M/s Wrecker Engineers (Germany)
    iv. Soil Investigations by M/s Indian Geotechnical Pvt. Ltd. and M/s Cengres.

(e) CPWD Engineers posted on this project did outstanding work and their role in execution of this project is highly appreciated the list of these Engineers is as given below. (a) Chief Engineer/Chief project Manager – (i) Jose Kurien, (ii) S K Rustagi and (iii) Dr. Shishir Bansal. (b) Superintending Engineer/ Project Manager – (i) Pradeep Garg, (ii) BK Sinha, (iii) Priyank Mittal and (iv) Vivek Bansal. (c) Executive Engineer – (i) BB Wadhwa, (ii) PK Sharma, (iii) AP Yadav, (iv) Suresh Pal and (v) RP Sharma, (vi) SB Gupta, (vii) SK Manchanda, (viii) Varun Kumar, (ix) Sunil Kumar Jain, (x) KC Pant, (xi) Ajay Kumar, (xii) KD Narayan, (xiii) Valsala Nair, (xiv) Nilabh Gupta, (xv) VK Gupta, (xvi) SA Khan, (xvii) BN Singh, (xviii) HS Rohilla, (xix) DD Bunker and (xx) NK Sarin. (d) Assistant Engineer – (i) Apoorva Kaushik, (ii) BL Kukreja, (iii) Vinod Garg, (iv) Soumen Bardhan, (v) LC Gupta, (vi) Davinder Singh, (vii) DK Gupta, (viii) Rakesh Kumar, (ix) Arun Kumar, (x) Karan Bir, (xi) Anoop Kumar, (xii) JP Chaurasia, (xiii) SK Sharma, (xiv) Satya Prakash, (xv) YD Sharma and (xvi) Saleem Marghoob.

(f) Engineers from the Construction Companies who worked for this project. They also did outstanding work - (i) VN Heggade as Project Director, (ii) Col (Retd) SG Diwanji as Chief Project Manager, (iii) R. Prakash as Addl. Chief Project Manager, (iv) Arvind Singh as Addl. Chief Project Manager, (v) Swapnil Nawalkar as Planning Engineer, (vi) Satbir Singh as Manager Execution of foundations and (vii) Gaurav Khanna as Manager Erection of Pylon.

(g) Grateful to Dr. Shishir Bansal for permitting to publish this abstract paper. The information has been compiled in this paper form a detailed paper by Shri Shishir Bansal published by The Institution Engineers in their Civil Engineering Journal 2021.

LESSONS LEARNT

(i) **Integration of Engineering with Art and Tourism:** Determinate resolution by Engineers and support as also confidence by DTTDC as also Govt. of Delhi ensured execution of this project. Consultancy services, material selection and execution for this project had many challenges as is seen from the above narrative. The project was implemented with success and our country should be proud of hard and sincere work by engineering professionals as also contractors.

(ii) **Adoption of new Technology and its challenges:** A confidence has developed in the country that new techniques and technologies can be implemented by our Engineers. They are equivalent to Engineers of any other advanced country. The success of this project should be propagated in our country and even in other countries. In fact suitable awards will encourage engineering professionals.

(iii) **Engineer's Pride:** Engineers from other parts of the country as also other countries should visit this project and those who were involved in the project should explain how they did it. It is much more than normal engineering projects.

Signature Bridge on River Yamuna, Delhi -
Magnificent view of the Signature Bridge.
On the upstream side existing bridge cum barrage is also seen

Signature Bridge on River Yamuna, Delhi -
Side view of the bridge tower and bridge

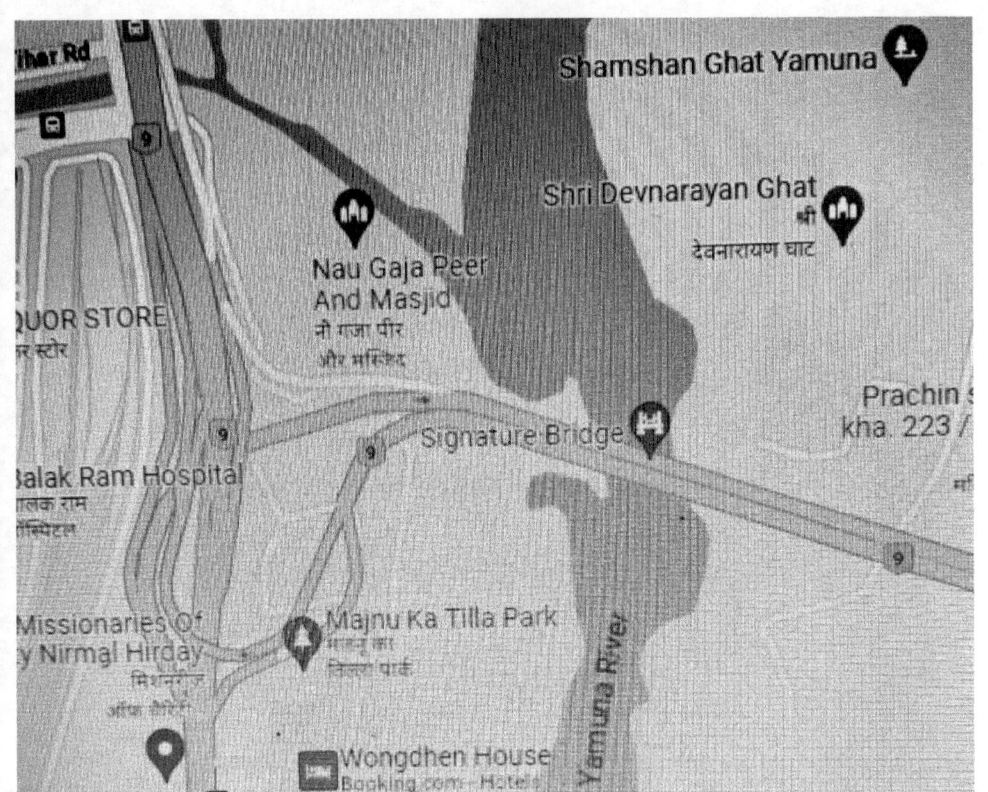

Signature Bridge on River Yamuna, Delhi –
Satellite map showing Signature Bridge and dispersal arrangement on Western end consisting of loops, ramps and flyovers

CASE STUDIES IN CONSTRUCTION PROJECT MANAGEMENT

KB RAJORIA & HK SRIVASTAVA

FEEDBACK FORM

. Which topics from your examination point of view are inadequately or not covered in this book?

____

____

. Is there any misprints/ mistakes/ factual inaccuracy in this book? If so, please specify.

____

____

. What is your assessment of this book as regards the presentation of the subject-matter, expression, precision and price in relation to the other recommended books available on this subject?

____

____

. Which competing books you regard as better than this? Please mention their author and publishers.

____

____

. Any further suggestions/comments you would like to make for the improvement of the book?

____

____

PS: If needed, please attach a separate sheet.

Name   : _______________________________

College   : _______________________________

Address : _______________________________

Email   : _______________________________

Mobile   : _______________________________

Please mail the feedback form at:

KHANNA BOOK PUBLISHING CO. (P) LTD.

4C/4344, Ansari Road, Daryaganj, New Delhi-110002

You can also email this feedback form at contact@khannabooks.com or

Whatsapp at +91 – 99109 09320

BEST FEEDBACK WILL BE REWARDED SUITABLY

www.ingramcontent.com/pod-product-compliance
Lightning Source LLC
LaVergne TN
LVHW022111210726

843510LV00015BA/1143